T0292069

CAMBRIDGE STUDIES
IN MATHEMATICAL BIOLOGY: 6

Editors
C. CANNINGS
Department of Probability and Statistics, University of Sheffield, UK

F. C. HOPPENSTEADT
*Systems Science and Engineering Research, Arizona State University,
Tempe, USA*

L. A. SEGEL
Weizman Institute of Science, Rehovot, Israel

AN INTRODUCTION TO THE MATHEMATICS OF NEURONS
SECOND EDITION

Modeling in the frequency domain

CAMBRIDGE STUDIES IN MATHEMATICAL BIOLOGY

FRANK C. HOPPENSTEADT

Arizona State University

An Introduction to the Mathematics of Neurons

SECOND EDITION

MODELING IN THE FREQUENCY DOMAIN

CAMBRIDGE
UNIVERSITY PRESS

CAMBRIDGE UNIVERSITY PRESS
Cambridge, New York, Melbourne, Madrid, Cape Town, Singapore, São Paulo

Cambridge University Press
The Edinburgh Building, Cambridge CB2 2RU, UK

Published in the United States of America by Cambridge University Press, New York

www.cambridge.org
Information on this title: www.cambridge.org/9780521590754

First published 1997

A catalogue record for this publication is available from the British Library

Library of Congress Cataloguing in Publication data
Hoppensteadt. F. C.
An Introduction to the mathematics of neurons : modeling in the
frequency domain / Frank C. Hoppensteadt. – 2nd ed.
p.   cm. – (Cambridge studies in mathematical biology : 14)
ISBN 0-521-59075-2 (hbk.). – ISBN 0-521-59929-6 (pbk.)
1. Neurons – Mathematical models.   2. Neural circuitry –
Mathematical models.   I. Title.   II. Series.
QP363.3.H67   1997
573.8′536′0151 – dc21   96-36822
                                                CIP

ISBN-13 978-0-521-59075-4 hardback
ISBN-10 0-521-59075-2 hardback

ISBN-13 978-0-521-59929-0 paperback
ISBN-10 0-521-59929-6 paperback

Transferred to digital printing 2006

*This book is dedicated to Leslie, Charles, Matthew,
Sarah, Robin and Sam*

Contents

Preface

... to such discontented pendulums as we are.

Ralph Waldo Emerson

Neurons, or nerve cells, control most biological rhythms through various timing mechanisms. Among these rhythms are body temperature and rest/activity behavior, but there is also significant modulation of timers due to external physical cues such as light/dark cycles and variations in ambient temperatures. Biological timers act on time scales ranging from milliseconds to months, and experiments on them range in size from microelectrode recordings to observations in underground laboratories that are used to study daily human rhythms over periods of months.

This book introduces some modeling techniques that are useful for studying rhythms and timing from the level of neurons to higher levels in the brain, and it focuses on the behavior of neurons and networks of them in the *frequency domain*. Although much of this material is based on the earlier work of many people who studied biological rhythms, including the first edition of this book published in 1986, much of it is new, reflecting knowledge about the brain that has been created in the past ten years [2,3]. For example, the study of networks has matured during this period, and we include new material on large networks. In addition, increases in knowledge about the brain and its subsystems make it possible to include here some special studies of attention, vision, and audition.

This second edition keeps to the same format as in the first, and it continues to focus on modeling in the frequency domain through the use of a particular model, the VCON (Voltage Controlled Oscillator Neuron) model. The VCON model was originally derived as being a good model for teaching and studying the phenomenon of phase locking in biological systems, but the mathematical model for it is quite similar to models that have proved to be useful in other settings:

- It appears as the canonical model for a general network near a multiple saddle-node on a limit cycle (SNLC) bifurcation.
- It occurs in describing emergence patterns of periodical cicadas [66].
- It arises as the model for a mechanical pendulum operating in an oscillatory or random environment [116, 135].
- It is used to describe nonlinear phenomena in power systems [73, 115].
- It arises as the model of a basic circuit used in communications theory (the phase-locked loop) [94].
- It is used to describe quantum mechanical aspects of superconductors, for example, Josephson junctions [91, 44].

In each of these cases, a mathematical model of a physical phenomenon is converted into phase and amplitude coordinates, and the model is analyzed using frequency-based methods of Fourier analysis and averaging.

Our goal here is to study the frequency and timing of neuron firing and how these can interact in networks to carry information and to control biological systems. We study the flow of information in large networks using methods similar to those used to study the flow of information in large telecommunications networks and the flow of alternating current electricity in power systems.

To study problems in the frequency domain, we work with angle (or phase) variables, and a common stumbling block for people entering this area is the question "What does the phase represent?" The following example might help get around this. Consider a one-handed stopwatch. Time is described by the hand's location, say whose tip has coordinates $(\cos 2\pi t/60, \sin 2\pi t/60)$ and t is usual time measured in seconds, relative to markings on the circumference. We can describe this timer by writing the location of the tip of the hand as being

$$(\cos \theta(t), \sin \theta(t))$$

and describing how the phase variable θ changes with t. In this case, θ is the solution of the differential equation

$$\dot{\theta} = \frac{2\pi}{60}.$$

In this way, we have moved the problem from the domain of physical variables (the position of the hand in this case) to the frequency domain ($\dot{\theta}$). Not much has apparently been gained by doing this. However, now consider a more complicated stopwatch where the markings around the edge are uneven to reflect systematic or random errors in the timer. This can be described using the original model but now the equation for θ must account for uneven progress of time on this clock, say

$$\dot{\theta} = \frac{2\pi}{60} + f(\theta),$$

where $f(\theta) > 0$ describes when the stopwatch time is moving faster than real time and $f(\theta) < 0$ describes when it moves slower than real time. Thus, the timing is modulated, and the modulation is conveniently described using a single equation for θ.

As another example, a voltage pulse might be described by a quite complicated function of time, say

$$V(t),$$

describing an action potential at a site on a nerve membrane. The voltage V lies in some interval, say

$$V_{\min} \leq V(t) \leq V_{\max}.$$

Such a function can frequently be described by a fixed (simpler) wave form having a variable phase. In these cases, a phase variable θ can be defined by

$$\cos \theta(t) = (V(t) - \bar{V})/\sigma,$$

where $\bar{V}$ is the mean of V:

$$\bar{V} = \frac{V_{\max} + V_{\min}}{2},$$

and σ measures its range:

$$\sigma = \frac{V_{\max} - V_{\min}}{2}.$$

The variable θ is a phase variable in the sense that it measures the extent of development of the signal, V. Typically, θ satisfies a differential equation that can be derived from the one for the physical variable V.

Phase variables in our applications eventually converge to the form

$$\theta \rightarrow \omega t + \phi,$$

where we call ω the *frequency* and ϕ the *phase deviation* of the signal. We will find that as in FM-radio, ω is like an address (on the radio dial) and ϕ carries timing information. Both ω and ϕ carry information. In high-dimensional cases (where ω, ϕ, etc. are vectors) such solutions form knots on a high-dimensional torus that phase locking shows will persist in the presence of noise [54].

Another important point of view presented and used here is that of bifurcations. Bifurcations often occur in systems and are observable in experiments.

These are circumstances where slight changes in data or state can result in dramatic changes in system behavior.

If a system is operating in a hyperbolic manner (there are no imaginary eigenvalues for relevant linearizations), then small changes in the system should not cause major disruptions in its behavior. In contrast, if some eigenvalues are near the imaginary axis, then a small change in the system's state or in the parameters describing it can result in dramatic changes in a network's behavior. Bifurcations can occur in many ways, but one of the more complicated elementary bifurcations to study is the fold bifurcation since it requires one to have nonlocal knowledge of the system. VCON models are quite useful for studying systems near such bifurcations since they are constructed on the basis of a saddle-node bifurcation near a limit cycle; consequently, they simultaneously capture a fold bifurcation and tractable nonlocal behavior. For example, consider the equation

$$\dot{\theta} = \omega + \cos\theta.$$

When $0 \le \omega \le 1, \theta \to \cos^{-1}\omega$ as $t \to \infty$. But, if $\omega > 1, \theta \to \infty$. If we are reading output of the system as a periodic function of θ, then the output stabilizes in the first case, and it oscillates in the second. This switch in the system's behavior from stable to oscillatory results from the saddle-node bifurcation that occurs when ω increases through the value $\omega = 1$.

Phase locking is important in electrical and mechanical systems [135]. It enables a system to provide a stable output even in the presence of significant levels of noise. Phase locking has also been found to occur in neural tissue [43, 46, 49, 5, 59, 60, 92, 87], and descriptions and analysis of phase locking are done in the frequency domain.

The emphasis here is placed on frequency and phase-deviation information of neurons. This is done by deriving a model in the frequency domain for the hillock region of a nerve cell where voltage pulses (action potentials) are triggered. Time delays in propagation of signals along membranes, across synapses, through dendritic trees and in cell bodies are modeled by including appropriate filters in the circuits. It is intriguing that frequency and timing of action potentials can result in the storage of information and in physiological and psychological responses of the system, and there remain many interesting and untouched aspects of this kind of information storage and retrieval and processing by neural networks.

How general is this approach using VCONs? Any model of a neuron or a network of them using dynamical systems will have the form

$$\dot{x}_i = \mathcal{F}_i(x), \tag{0.1}$$

where $i \in \overline{1, N}$ (where N is a large number) describes the addresses of all elements in the network, and the vectors x_i, $\mathcal{F}_i \in E^{m_i}$ describe the dynamics of each circuit element and the impact of all other elements on it.

This system can change from static to oscillatory behavior in two simple (codimension = 1) ways: Either through a saddle-node on a limit cycle (SNLC) bifurcation or through a Hopf bifurcation. Both of these are observed in neuroscience experiments. There are many other ways that this can happen, but they involve more constraints (higher codimension) and in that sense are less likely [58]. In the SNLC case, there is an invariant limit cycle that is homeomorphic to a circle, say $x = C(s)$ where $s \in \overline{0, 2\pi}$ such that $|C'(s)|^2 > 0$ for all s, $\mathcal{G}(s) \equiv C'(s) \cdot \mathcal{F}(C(s))/|C'(s)|^2$ satisfies $\mathcal{G}(0) = \mathcal{G}'(0) = 0, \mathcal{G}''(0) > 0$ and $\mathcal{G}(s) > 0$ for $0 < s < 2\pi$. In addition, $N - 1$ of the eigenvalues of $\mathcal{F}_x(C(s))$ have nonzero real parts for each $s \in \overline{0, 2\pi}$. Then the system restricted to C is simply $\dot{s} = \mathcal{G}(s)$, and its canonical model is $\dot{\theta} = \omega + \cos\theta$ for ω near 1 in the sense that there is a smooth invertible function $h : S^1 \rightarrow S^1$ such that for any solution for s, there is a solution for θ such that $s(t) \equiv h(\theta(t))$. The equation for θ here is the core of the VCON models.

The Hopf bifurcation case is studied elsewhere. It is not studied in depth here because phase locking methods for it require more technical mathematics to derive than our approach here.

We focus here on VCONs and networks of them to gain insight to how such systems, in particular ones that encounter saddle-node bifurcations, can process information.

The first chapter introduces elementary circuit theory and some of its mathematics. Particularly important in this chapter is the introduction of voltage-controlled oscillators (VCOs) and some elementary circuits that use them. VCOs are the central devices in our frequency-based neuron theory developed in later chapters.

Chapter 2 discusses some mathematical aspects of clocks. In particular, it is shown that VCO circuits are quite similar to *simple clocks*, which have helped our understanding of biological rhythms. Phase-resetting experiments are also described in Chapter 2. Finally, it is shown how simple clocks are related to neurons. This chapter is intended to introduce ideas of modeling and analysis in the frequency domain.

The third chapter describes the physiology of neurons and some electrical circuit analogs of them. Among the latter are the Hodgkin–Huxley model, the FitzHugh–Nagumo model, and some simplifications of them. These represent the traditional approach to neuron modeling. The VCON is also introduced in Chapter 3.

The VCON model is quite similar to relaxation oscillation models of neuron behavior but, unlike relaxation oscillators, it is surprisingly simple to study since it accurately models phase aspects of the circuit while avoiding technical asymptotic approximations. For example, the most sophisticated model of neurons to date involves variables that account for ionic currents and the opening and closing of channels for them, but these are derived in terms of physical quantities of voltage, current, and chemical concentrations. Study of such models for phase locking requires conversion of these variables, using mathematical methods, to phase and amplitude variables, which may not be possible. On the other hand, the model of a VCO is posed in phase variables from the start, by design of some brilliant electrical engineers, to facilitate direct study without use of technical mathematical transformations.

Chapter 4 deals with signal processing in phase-locked feedback circuits, and it sets the scene for our later treatment of signal processing in neural networks. Phase-locked loops (PLLs) are analyzed by using the rotation vector method. It is shown in Chapter 4 that the VCON is a PLL, and the rotation vector method is then applied to describing the stable response of a VCON to external oscillatory forcing. This approach shows how to construct an energy-like function, the minima of which correspond to the stable responses of the VCON.

Several examples of small neural networks are modeled and analyzed in Chapter 5. Among these are a simple bursting pattern generator (the Atoll model), the control of respiration during exercise, and the mechanisms of rhythm splitting in crepuscular mammal activity. Numerical simulations and the rotation vector method are used to determine phase-locking behavior within these small networks.

Large neural networks are studied in Chapters 6 and 7. Chapter 6 describes memory, phase changes, and synchronization in networks; Chapter 7 describes certain networks in and near the neocortex. These networks respond to external stimulation in a variety of interesting and complex ways. Energy surfaces have been derived by others to clarify some responses of networks of On-Off neurons, or Ising-like networks, and Chapter 6 describes some of their work. However, we consider here more realistic problems where the equations we derive are gradient-like fields for phase deviations between synchronized oscillators. This associates with stable phase deviations a memory surface, or as we say here, a *mnemonic surface*. The mnemonic surface approach allows us to interpret stable firing patterns of parts of the brain as representing its short-term memory and its behavioral response. VCON networks are very rich in stable firing patterns and, although they are not as complicated as ionic current circuits, they clearly show how firing frequencies within the network can store, recall, and process phase information. Energy-like surfaces have been used by psychologists in interesting ways: The work of Helmholtz, Freud, and Jung illustrates their impact.

Chapter 7 is largely devoted to creating and studying networks that model parts of the neocortex and its collateral processes. We aim at the thalamic searchlight, a paradigm for focusing attention, and at pattern formation and wave propagation in the neocortex and the visual cortex.

The appendixes at the end of the book present certain methods of differential equations that I have found to be useful review for students with varying backgrounds in mathematics and a brief summary of topics from bifurcation theory that are relevant to this work. Recommended for further reading are [11, 36, 72, 92].

There have been a number of developments since publication of the first edition of this book that bear directly on the present book. In fact, they constitute the reasons for pursuing a second edition. The following list of items is now folded into the present edition:

- The emphasis of this work on the frequency domain and the signal processing methods developed here are highlighted and expressed in terminology that is closer to common usage in the engineering literature.

- Additional material is included on how noise in signals affects the models derived here. New material is included describing the works of Skorokhod, Chetaev, Kuramoto, Wentzel, and Friedlin in the context of neuroscience.

- More work is included on both chemical and electrical synapses. For the most part the analysis in the first edition was for electrical synapses only.

- New work is presented relating the approach taken in this book with the rest of the neuroscience modeling literature, much of which has emerged since the publication of the first edition. Our approach is through the frequency domain, and new material is included that describes how ionic channel models can be converted for study in the frequency domain and how variables in the frequency domain models are related to physiological variables. In particular, we formulate the VCON model in terms of activity and phase where activity is interpreted as being the firing rate of a cell and phase as having (eventually) the form $\omega t + \phi$.

- Material on bifurcations is now included. The VCON model developed and studied here turns out to be closely related to the canonical model of a fold (or saddle-node) bifurcation. This connection between VCONs and bifurcations is interesting since most bifurcation phenomena, or phase changes in a system, are detectable in experiments. Therefore, it is through connections between bifurcations in the model and predicted phase changes in a network that we can relate our model to possible experiments. Bifurcations involve dimensionless parameters that should be accessible through experiments.

- There are numerous tantalizing connections between the VCON model and models from quantum mechanics. This might be relevant to studies of consciousness that are now ongoing [107], and some of these similarities are pointed out in the present edition.

• Several new simulations have been developed and carried out for neural circuits using VCON methodologies. These include

   – Attention networks in the thalamus–reticular complex region of the brain.
   – Shivering and flight governed by central pattern generators in moths.
   – The development of mnemonic surfaces, that is, surfaces that describe remembered states of a system.
   – The dynamics of cortical columns. We derive and study the *pencil model* in Chapter 7.
   – The development of ocular dominance in the visual cortex of newborns. This is modeled using synapse strengthening due to synchronous firing of presynaptic and postsynaptic cells.

The work presented in this book is based on courses in mathematical modeling where this material has been used. It can do no more than provide a snapshot of certain aspects of the brain in a research field that is creating masses of new and useful material every day. The focus here is on mathematical modeling, and thus this material should not be construed as being useful directly for clinical purposes.

Acknowledgments

I thank Tatyana and Eugene Izhikevich for help in the production of this second edition. In addition, I thank my many colleagues and students who have contributed to this book through helpful comments and critical discussion of material in it. In particular, work by Humberto Carrillo, Catherine Garcia, David Pettigrew, Brian Platt, Jeffrey Schall, and Diana Woodward bears directly on topics presented here.

1

Some useful electrical circuits

Electrical circuits are important in our lives, certainly in electric guitars, television sets, and computers. However, they are also important in understanding how our bodies work. In fact, most nerve activity, including that in the brain, is electrical: Ionic currents passing through membranes and across synaptic gaps are the dominant physical properties of neurons. This first chapter introduces some basic elements of electrical circuits, starting with simple resistors and ending with a description of some modern integrated circuit chips that behave like neurons.

Electrical circuits are described in terms of the physical quantities of voltage (V) and current (I). Voltages and currents are not intuitive; they cannot be directly observed by us without special instruments such as voltmeters and ammeters. However, we can think of voltage as being a pressure that pushes electrons in a conductor and of current as measuring the electron flow. We gain intuition about voltage and current by thinking of them as being solutions of the appropriate mathematical models, either ones derived here from Kirchhoff's Laws for the elementary circuits or, more generally, ones derived from Maxwell's equations in more advanced work.

The circuits studied here involve several components or *circuit elements*. These are listed next along with their notations and *IV* (current-voltage)-characteristics. Circuit elements can be combined to form circuits, and these can be modeled using Kirchhoff's laws. *RLC* circuits are important examples of simple circuits. Other important circuits presented in this chapter include filters and oscillator feedback circuits.

1.1 Circuit elements

Circuits are combinations of physical devices such as resistors and capacitors, and they are described using mathematical terms.

1

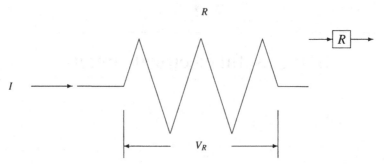

Figure 1.1. A resistor. A current I passing through the resistor with resistance R creates a voltage change across the device of size $V_R = RI$.

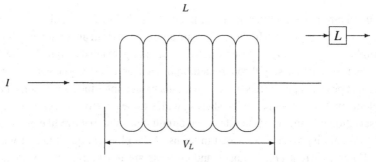

Figure 1.2. An inductor. A current I passing through an inductor creates a voltage change of size $V_L = L\dot{I}$.

Resistors are devices that impede the flow of current. Let I be the current into the resistor and V_R the voltage across it as shown in Figure 1.1. Observations of how V_R and I are related led to *Ohm's law*

$$V_R = RI.$$

That is, the voltage across a resistor is proportional to the current through it. The constant of proportionality, the *resistance R*, is measured in units of ohms. A high (low) resistance with a fixed voltage results in a low (high) current.

Inductors are coils of wire wrapped around a metal core (see Figure 1.2). Current through the coil induces a magnetic field in the core that creates a voltage. In an inductor I and V_L are related by the formula

$$V_L = L\dot{I},$$

where $\dot{I} = dI/dt$. The constant L is called the *inductance*, and it is measured in units of henrys. Inductors are important in circuits described in later chapters,

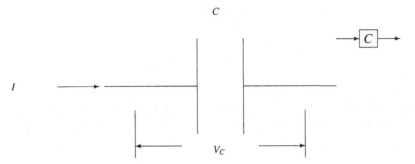

Figure 1.3. A capacitor. Charge accumulates on the plates of the capacitor at a rate proportional to I, and there results a voltage change $\dot{V}_C = I/C$.

but there they are replaced by more convenient integrated circuits. Still, the result has the same *IV*-characteristic and so inductors are written into circuits even though they are replaced by other devices.

A *capacitor* is a device that accumulates charge on plates separated by a nonconductor (Figure 1.3). The charge coming in (I) accumulates and the *IV*-relation is

$$V_C = \frac{1}{C} \int_0^t I \, dt$$

or, equivalently,

$$I = C\dot{V}_C.$$

The constant C is called the *capacitance*, and it is measured in units of farads. In most of the circuits used here the appropriate units are microfarads (10^{-6} farads).

1.1.1 Electromotive force

A power supply, such as a battery or an alternating voltage, applies an electromotive force (voltage) to a circuit. We denote an electromotive force by E and depict it as shown in Figure 1.4. When arrows are present, they indicate the direction from positive to negative for the variables they describe.

1.1.2 Voltage adders and multipliers

Voltages can be added and multiplied by using a combination of operational amplifiers [68]. Figure 1.5 indicates the notation used for a voltage adder and

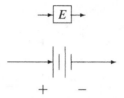

Figure 1.4. A battery. Although it is indicated here, we ignore the polarity of batteries in our subsequent diagrams, but this is corrected for in the signs of currents and voltages in the circuits.

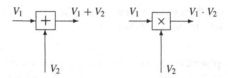

Figure 1.5. A voltage adder (left) and a voltage multiplier (right).

for a voltage multiplier. The result of input voltages u and v is their sum $u + v$ or their product $u \cdot v$, respectively.

1.2 Filters

An important class of circuits involves a resistor, an inductor, and a capacitor in series with a battery. These are called *RLC* circuits. A mathematical description of an *RLC* circuit can be derived using Kirchhoff's laws.

Filters are important since they sort out and allow to pass only certain frequencies. The purpose is often to eliminate noise from a signal or to restrict a signal to a size that meets tolerances of circuit elements farther downstream. The filters described here are used in various applications later.

1.2.1 Kirchhoff's laws

Most circuit models are derived using Kirchhoff's laws. These state that:

- The total voltage measured around any closed loop that can be drawn in the circuit is zero.
- The total current into any circuit node sums to zero.

A circuit node is any point at which two or more wires come together, and a closed loop in a circuit is any closed loop that can be made on a circuit diagram. These definitions are illustrated in the next section.

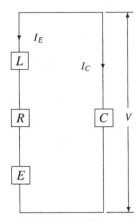

Figure 1.6. An *RLC* circuit.

1.2.2 RLC circuits

An *RLC* circuit is shown in Figure 1.6.

The first of Kirchhoff's laws implies that $I \equiv I_E$ satisfies

$$-E + RI + L\dot{I} + V = 0,$$

where the four terms on the left are the applied voltage (measured in the clockwise direction) and the voltages across the resistor, inductor, and capacitor, respectively. The current and the capacitor voltage are related by

$$C\dot{V} = I.$$

Thus, we get a system of two differential equations:

$$C\dot{V} = I,$$

$$L\dot{I} = E - V - RI.$$

These equations can be solved in closed form for V and I once C, L, R, and E are known. This is performed a little later, but first we will consider the geometry of solutions.

1.2.2.1 Geometry of solution of an RLC circuit

Geometric methods play a big role in the study of *RLC* circuits and more complicated circuits. For this, a plot of I versus V is made, and graphs of the solutions to this system of equations are constructed.

The first step is to find *isoclines*. These are the graphs on which $\dot{V} = 0$ and on which $\dot{I} = 0$. These are depicted in Figure 1.7.

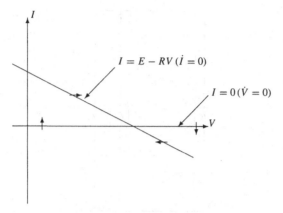

Figure 1.7. Isoclines of the *RLC* circuit.

Obviously, the static states, or equilibria, of the system occur where the isoclines cross: In this case, there is one equilibrium ($I = 0$ and $V = E$). Since ($\dot{V}, \dot{I}$) is a vector tangent to the solutions, we can roughly draw the solutions by observing how they cross the isoclines and axes. Typical crossings are shown in Figure 1.7. It follows that solutions oscillate around the equilibrium.

Next, we determine the solutions in closed form. This is possible because the *RLC* circuit is a linear circuit (see Appendix A).

1.2.2.2 Analytic solution of an RLC circuit

Differentiating the first equation leads to a single second-order differential equation

$$LC\ddot{V} + RC\dot{V} + V = E.$$

This equation can be solved using the Laplace transform method, as shown in Appendix A. Note that when resistance in the circuit is negligible, the model reduces to a harmonic oscillator

$$LC\ddot{V} + V = E,$$

where the natural frequency is shown in Appendix A to be $\sqrt{1/LC}$.

The ratio R/L represents the damping in the full circuit. From the Appendix, we see that if $R > 0$, then solutions spin into the equilibrium. If $R = 0$, then the solutions are ellipses about the equilibrium.

1.2.2.3 LC circuits

RLC circuits with no resistance behave like timers. The circuit shown in Figure 1.8 is referred to as an *LC* circuit.

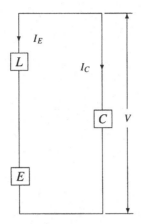

Figure 1.8. A harmonic oscillator.

As shown next, the solutions of this circuit for V and I describe ellipses in the V-I plane centered at $V = E$, $I = 0$. A radius drawn from this center to the point $(V(t), I(t))$ moves like one of the hands on a clock. In fact, the LC circuit can easily be converted to the frequency domain: Let

$$V = E + r \sin \theta$$

and

$$\sqrt{LC}\dot{V} = r \cos \theta.$$

After some calculation, we obtain

$$\tau \dot{r} = 0$$

$$\tau \dot{\theta} r = r,$$

where the time constant $\tau = \sqrt{LC}$. Therefore, $r = $ constant. If $r \neq 0$, then $\dot{\theta} = 1/\tau$, and equation in the frequency domain completely describes the voltage and current dynamics in the circuit: $V = E + r \sin t/\tau$ and $I = C\dot{V} = r\sqrt{C/L} \cos t/\tau$.

1.2.3 RC circuits; low-pass filters

The circuit in Figure 1.9 describes a low-pass filter where $E = V_{\text{in}}$ is the input voltage and $V = V_{\text{out}}$ is the output voltage.

The mathematical model of a low-pass filter is

$$-V_{\text{in}} + RI + V_{\text{out}} = 0,$$

$$C\dot{V}_{\text{out}} = I.$$

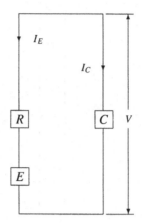

Figure 1.9. A low-pass filter.

This leads to the single equation

$$RC\dot{V}_{\text{out}} + V_{\text{out}} = V_{\text{in}}.$$

Given V_{in}, we can solve this equation to get

$$V_{\text{out}}(t) = V_{\text{out}}(0)e^{-t/RC} + \frac{1}{RC}\int_0^t e^{-(t-s)/RC}V_{\text{in}}(s)\,ds.$$

The last term in this formula is called a convolution integral; in it, past values of V_{in} are weighted and added up.

1.2.4 Transfer functions

An input–output device where the input voltage W is related to the output voltage V by the equation

$$a_n V^{[n]} + a_{n-1}V^{[n-1]} + \cdots + a_0 V = b_m W^{[m]} + \cdots + b_0 W,$$

where $V^{[j]} = d^j V/dt^j$, etc., is referred to as a *general filter*. This equation can be solved using Laplace transforms, as shown in Appendix A. Each derivative is replaced by the transform variable, say s, to the appropriate power: The result is that if $\widetilde{V}(s)$ and $\widetilde{W}(s)$ denote the Laplace transforms of V and W, respectively, then

$$(a_n s^n + a_{n-1}s^{n-1} + \cdots + a_0)\widetilde{V} = (b_m s^m + \cdots + b_0)\widetilde{W} + \text{a polynomial in } s.$$

The last polynomial in s depends on initial conditions, and it is taken to be known from initial conditions. Therefore, the transform of the output voltage is found to be a rational function of s multiplied by $\tilde{W}$:

$$\tilde{V}(s) = \tilde{W}(s)[b_m s^m + \cdots + b_0]/[a_n s^n + \cdots + a_0],$$

where terms involving the initial conditions are ignored. The inverse transform formula gives a convolution integral for V:

$$V(t) = \int_0^t h(t - t')W(t')\,dt' + \text{ a known function},$$

where h is a function whose Laplace transform is

$$\tilde{h}(s) = [b_m s^m + \cdots + b_0]/[a_n s^n + \cdots + a_0].$$

The function $\tilde{h}$ is called the *transfer function* of the filter, and it is convenient to follow the engineering literature and write

$$V(t) = \tilde{h}(s)W(t)$$

for the convolution integral formula.

Using this notation, we can write the input–output relation for a low-pass filter as

$$V_{out}(t) = \frac{1}{RCs + 1} V_{in}(t),$$

which carries all the meaning derived earlier in this section. In particular, this is a convenient shorthand notation for the integral applied to the input voltage.

1.3 Voltage-controlled oscillators (VCOs)

Voltage-controlled oscillators are the basic elements used in some neuron models studied in later chapters. They are electrical oscillators whose frequency is modulated or controlled by an input voltage. There are many kinds of voltage-controlled oscillators available on the electronics market, but we denote a generic one by VCO and depict it by the graph

$$V_{in} \to \text{VCO} \to V(x(t)),$$

where the input voltage V_{in} and the output voltage V are related in a somewhat complicated way. V is a fixed function, called the wave form (e.g., a square wave, a saw-tooth wave, or a sinusoidal wave). That is, it is a function that is periodic in its argument x and has a fixed shape. All three of these possibilities are available directly from outputs on commercially available VCOs. The function $x(t)$ is the phase of the output voltage. For example, if we select the sinusoidal wave output (a certain pin coming out of the chip) and the device has settled down under steady conditions, then $x(t) = \omega t + \phi$, where ω is the output frequency, ϕ is the phase lead (if positive) or phase lag (if negative), and the output voltage would be $\sin(\omega t + \phi)$.

There exists some confusion in the terminology since sometimes ϕ is referred to as the phase of the output; we will try to be consistent here and describe x as being the phase of $V(x)$ and when appropriate ϕ will be the *phase deviation*.

For example, the potential supplied to households in the United States is 117 volts (root mean square) with a 60-cycle alternating current, and the voltage observed across the terminals in a wall socket is (approximately) $165 \cos(2\pi \frac{t}{60} + \phi)$, where time is measured in seconds. The amplitude of the voltage is 165 and its phase is $2\pi \frac{t}{60} + \phi$ (with $\phi = 0$, $2\pi/3$, or $4\pi/3$). Here $V(x) = \cos x$ and $x(t) = 2\pi \frac{t}{60} + \phi$. The phase difference, say $\phi_1 - \phi_2$ between two wires in a household, determines the size of the potential drop between the wires.

Current is ignored in VCOs, and the model is given in terms of the input and output voltages alone. So rather than using an *IV*-relation as we did for filters, we describe the device in terms of an input–output relation. The output of the VCO is an oscillatory function V of the phase $x(t)$. When V_{in} is in the operating range of the VCO, the output phase is related to this controlling voltage by the simple differential equation

$$\dot{x} = \omega + \sigma V_{in},$$

where ω is called the VCO *center frequency* and σ is the VCO sensitivity, both of which are known. Here and below, we take $\sigma = 1$ by suitably scaling input voltages. Keep in mind that the units in this equation are correct although σ does not appear explicitly in the following models.

This equation can be solved by integrating it:

$$x(t) = x(0) + \omega t + \int_0^t V_{in}(s)\,ds,$$

where $x(0)$ is the initial phase. The output voltage is

$$V\left(\omega t + \int_0^t V_{in}(s)\,ds + x(0)\right).$$

Thus, the larger is ω or V_{in}, the faster V will oscillate.

We suppose here that all voltages are within the operating range of the VCO device.

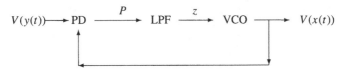

Figure 1.10. A second-order phase-locked loop.

1.4 Phase-locked loops (PLLs)

VCO feedback loops are very important in electronic communications, for example, in cellular telephones and computer terminal modems. The basic circuit is shown in Figure 1.10. The circuit depicted describes a second-order phase-locked loop. The circuit elements are a VCO, a low-pass filter (LPF), and a phase detector (PD). The low-pass filter output voltage is denoted by z and the phase detector output voltage is denoted by P. The input voltage and the output voltage are assumed to have the same wave forms (V), and y denotes the input phase and x the VCO phase.

The phase detector output depends upon which among many possible choices for this circuit element is used. For example, the wave form V is often sinusoidal and the phase detector output is often simply the product of its inputs. The result is a sum of sinusoidal functions of sums and differences of the phases. The LPF is tuned so that only the terms involving differences of phases pass. A typical case for P is $P = K \sin(y(t) - x(t))$, where K is called the *phase detector gain*. With this, and our work earlier in this chapter, we derive the following mathematical model for the circuit:

$$\tau \dot{z} + z = K \sin(y - x),$$

$$\dot{x} = \omega + z.$$

In this model, the filter time constant (τ), the VCO center frequency (ω), and the input phase ($y(t)$) are given and the equations describe how the output phase ($x(t)$) is related to them. Eliminating z from this model gives

$$\tau \ddot{x} + \dot{x} - K \sin(y - x) = \omega.$$

This equation is not easy to solve, and we must rely on other methods to discover how its solutions behave.

Another choice for a phase detector is a simple voltage adder where

$$P = \cos y + \cos x.$$

Although this may seem too simple to be useful, we must keep in mind that trigonometric functions have some unexpected features. For example, in this

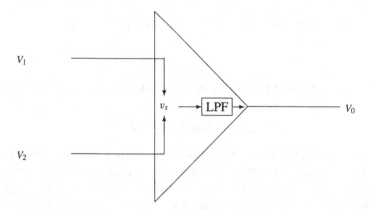

Figure 1.11. An operational amplifier.

case we have

$$\cos y + \cos x = \cos\left[(x+y)/2\right]\cos\left[(x-y)/2\right].$$

This identity will play an important role later.

1.4.1 First-order PLLs

Our understanding of PLLs is helped by considering the PLL without the presence of the low-pass filter. The resulting circuit is referred to as being a first-order PLL, and its mathematical description can be found simply by setting $\tau = 0$. The result is

$$\dot{x} = \omega + K \sin(y - x).$$

This equation is significantly easier to solve, and we will investigate its solutions later. Of course, in removing the low-pass filter, we remove some important features of the full PLL; for example, noise suppression is reduced.

1.5 Operational amplifiers (OpAmps)

An operational amplifier (OpAmp) is a circuit element like that depicted in Figure 1.11.

V_0 is the output voltage, and V_1 and V_2 are input voltages. Roughly speaking, the element acts by taking the difference between V_2 and V_1, passing it through an element whose output is v_s, and filtering this output, with the result being V_0.

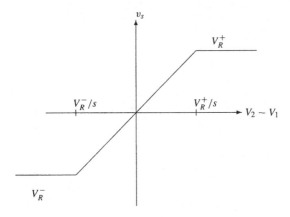

Figure 1.12. v_s as a function of $V_2 - V_1$.

The voltage v_s is depicted in Figure 1.12. In rough terms, the device's output tries to follow the larger of the inputs.

$V_R^{\pm}$ are the rail voltages (i.e., they give the maximum outputs of the device), and s, the slope of the diagonal line, measures the inaccuracy of the device. In theory, $s = \infty$. It follows that the output is determined from the equation

$$RC\dot{V}_0 + V_0 = v_s[V_2 - V_1].$$

The term C represents parasitic capacitance, and RC should be viewed as being very small. Let us take $RC = 0$. Then

$$V_0 = \begin{cases} V_R^+ & \text{if } V_2 - V_1 > V_R^+/s, \\ V_R^- & \text{if } V_2 - V_1 < V_R^-/s, \\ s(V_2 - V_1) & \text{otherwise.} \end{cases}$$

OpAmps are always used with feedback. Two examples are the linear amplifier and the voltage tracker.

1.5.1 A linear amplifier

The circuit shown in Figure 1.13 acts like a linear amplifier.

To see this, note that

$$(V_0 - V_1)/R_1 = V_1/R_2.$$

Therefore,

$$V_1 = rV_0,$$

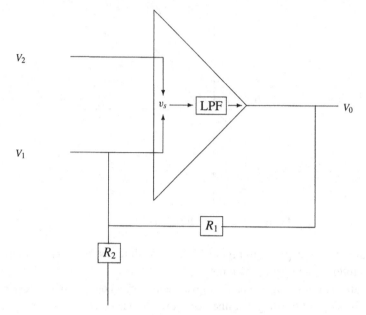

Figure 1.13. A linear amplifier.

where

$$r = R_2/(R_1 + R_2).$$

Substituting these values into the expression for V_0, we get

$$V_0 = \begin{cases} V_R^+ & \text{if } V_2 - rV_R^+ > V_R^+/s, \\ V_R^- & \text{if } V_2 - rV_R^+ < V_R^-/s, \\ s(V_2 - rV_0) & \text{otherwise.} \end{cases}$$

Ideally, $s = \infty$. Thus, setting this value for s gives

$$V_0 = \begin{cases} V_R^+ & \text{if } V_2 > rV_R^+, \\ V_R^- & \text{if } V_2 < rV_R^+, \end{cases}$$

If we define $V_0 = h(V_2)$ by this table of values, then the graph of h is shown in Figure 1.14.

Since $r < 1$, this circuit acts like a linear amplifier in the range

$$rV_R^- < V_2 < rV_R^+.$$

Beyond that range, it puts out the rail voltages.

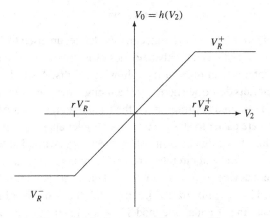

Figure 1.14. Output of a linear amplifier.

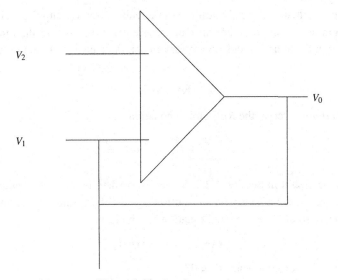

Figure 1.15. A voltage tracker.

1.5.2 A voltage tracker

The feedback circuit depicted in Figure 1.15 is called a voltage tracker since

$$V_0 = \begin{cases} V_R^+ & \text{if } V_2 > V_R^+, \\ V_R^- & \text{if } V_2 < V_R^-, \\ V_2 & \text{otherwise.} \end{cases}$$

This table of values follows from the linear amplifier calculation.

1.6 Summary

The elements of circuit theory presented in this chapter are used throughout the
later chapters to describe electrical analogues of neurons. In particular, VCOs
play a fundamental role in most of the following results. A striking feature of
the integrated circuits described here is that although they are quite complicated
combinations of hundreds of transistors, their input–output relations are quite
simple, and they are easier to study than some simpler appearing circuits! This
of course was done by design by brilliant engineers who wished to facilitate the
design of more complicated circuits. The filter circuits and feedback loops illustrate how these devices can be combined and modeled. Note that Kirchhoff's
laws are not used in these integrated circuit models, but only their input–output
relations are used. In particular, the models are not in terms of voltages, but in
terms of phases, which makes things simpler for mathematical analysis in the
frequency domain.

It is not difficult to start building these circuits. A power supply, a circuit
board, and an oscilloscope are all that are needed to begin, and the circuits
described in Sections 1.4 and 1.5 are easy to obtain, to construct, and to study.

1.7 Exercises

1. *RLC circuit.* Derive the *RLC* circuit equation

$$LC\frac{d^2V}{dt^2} + RC\frac{dV}{dt} + V = E$$

   from the circuit in Section 1.2.2. Solve this by determining what values of
   r are needed to make $\exp(rt)$ a solution. Denote these values by r_1 and r_2.
   Then any solution of the *RLC* equation has the form

$$A_1 \exp(r_1 t) + A_2 \exp(r_2 t).$$

   What is the solution when $R = 0$?

2. *Harmonic oscillator.* Show that if $R = 0$ in an *RLC* circuit, then

$$LI^2 + C(V - E)^2 = constant.$$

   (Hint: Show that

$$\frac{d}{dt}\left(LI^2 + C(V - E)^2\right) = 0.)$$

   Deduce that solutions describe ellipses in the *V-I* plane.

3. *Low-pass filter solutions.* Derive the formula for $V_{out}(t)$ in the low-pass filter
   described in Section 1.2.3.

4. *Response ratio.* Apply an input voltage that has frequency ω to a low-pass filter. What is the amplitude of the response? (Hint: Let $V_{in} = \exp{(i\omega t)}$ and solve for $V_{out}(t)$.) Use $V_{out}(0) = 0$. The response ratio is defined by the limit

$$\lim_{t\to\infty} \frac{|V_{out}(t)|}{|V_{in}(t)|}.$$

Evaluate this ratio for large t.

Plot the response of the circuit against the input frequency ω, and show that only the low frequencies are detectable in the output.

5. *Transfer function.* Derive the transfer function for a low-pass filter.
6. *A damped pendulum.* Sketch solutions of the equation

$$\frac{d^2x}{dt^2} + \frac{dx}{dt} - A\cos x = 0$$

in the $(x, dx/dt)$-phase plane.

2

A theory of simple clocks

We grow up with clocks, but we do not think much about them. We learn to tell time by the location of the long and short hands on a clock face, but we ignore the fact that the numbering system is discontinuous. We also expect that the hands proceed at a uniform rate throughout the day.

There are many timing devices in our lives that distort these features. Body timers such as temperature or blood hormone levels change rapidly through parts of a day or month and slowly through other parts. A similar effect on a wall clock would be to accelerate the hands during the morning and to slow them down during the afternoon. Or, we could take a 24-hour clock and use unequal spacings of the marks around the edge to tell time, with maybe morning times being spaced closer and afternoon ones farther apart. When we speak of faster or slower motion, we are speaking about the motion of the hands relative to some reference timer, such as solar time.

A mechanical clock has three main ingredients:

1. An escapement, which is a mechanism that sets a regular beat. The escapement might be driven mechanically by a pendulum or spring or electrically by an alternating current.
2. A transmission or modulator. This is a time scaling device or reduction mechanism that translates the escapement's beat into a movement useful for observation.
3. A clock face, where hands (or numbers) driven by the modulated beat depict time usefully.

Biological clocks have the same ingredients. In the ones studied here, neurons set the beat like a pendulum does for a mechanical clock; filters and actuators, like muscles or glands, scale the time by averaging neuron output; and a physiological function, like regular respiration or regular hormone cycles, represents

18

the output. Our approach is to show how a clock face can be embedded in a physiological process and so interpret the process as being a timer. To do this, the process must be described in terms that can be identified as being a mathematical clock, and the result is a conversion of the process from being described by physical variables of voltages, currents, etc. to the frequency domain where variables are phases and amplitudes.

This chapter begins with a simple clock, like one on the wall, and we derive a mathematical model of it. Simple clocks are related to basic building blocks of the neural networks that we consider in later chapters.

Interacting clocks modulate each other in many interesting ways. This was reported by Huygens in 1673 [71] for pendulum-driven clocks mounted on a common wall and by Galileo for chandeliers in a cathedral. Therefore, it is important to know how a simple clock can be modulated by outside forces and how modulation can be accounted for mathematically. We do this next.

Clocks defined by physiological processes are not simple clocks, but are more general and intriguing timers. The idea of a rubber-handed clock helps to describe how these more general clocks work. The rubber-handed clock captures some interesting and surprising timing anomalies.

Phase-resetting experiments provide excellent opportunities to investigate physiological clocks in a laboratory. In these, the regular activity of an organism may be studied: Exercise-wheel activity of hamsters is a good example. In such an experiment, the times when the animal uses its exercise wheel are recorded over several weeks under normal lighting conditions. A regular nocturnal pattern of use is observed. Then the animals are moved to a completely dark environment. Regular wheel activity is observed, but its period drifts to less than 24 hours. The phase-resetting experiment consists of shining bright light on the animal for a short time, say during its activity period, and recording what effect this stimulus has on wheel use. Some results of these experiments are described in Chapter 5, but in this chapter, these experiments are performed on the rubber-handed clock and the VCON to lay a basis for later comparisons.

Finally, we show how a rubber-handed clock can be found in nerve models and under what circumstances the rubber-handed clock can be approximated by a simple clock.

2.1 Some clock models

Phase-resetting experiments are important because they can be carried out in a laboratory without damaging the subject. It is therefore essential to keep the real experiments in mind when developing a theory for them so that the formulation is in line with that used by experimenters.

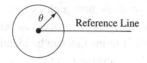

Figure 2.1. The rubber-handed clock.

Phase resetting can be visualized using a simple device called a rubber-handed clock. This is introduced in this section, and it is followed by a mathematical description of how simple clocks can be modulated.

2.1.1 The rubber-handed clock

Many timing devices in the body behave like a rubber-handed clock. Figure 2.1 depicts a clock face that has a single hand.

We suppose that the hand has curious physical properties: If it is stretched or compressed and let go, then it returns immediately to its original length, but length in the direction in which it was released; except if the hand is crushed to the center, then it stays there. The phase-resetting experiments are like stretching the hand of a rubber-handed clock in various directions, letting it go, and observing what change there is on the timer. See Figure 2.2.

Clocks of this kind will be found throughout this chapter, and understanding the concepts involved with them will help in following other material in later chapters.

Some further mathematical aspects of simple clocks must be developed before we proceed with a study of phase-resetting experiments for rubber-handed clocks.

2.1.2 Modulating simple clocks

Time can be measured according to an arbitrary scale around the edge of a clock face: Minutes, as on a wrist watch or radians, as shown in Figure 2.1. We choose radians for mathematical convenience. If θ denotes the angle (i.e., time) shown on the clock, then θ might increase in proportion to solar time, or it might change at some other rate. For example, we could set $\theta = \alpha t$, where t indicates Greenwich Mean Time, say measured in seconds, and α denotes the number of radians moved on the clock face per second. It is more convenient to write this description of the clock as a differential equation

$$\dot{\theta} = \alpha,$$

which indicates that the rate of change of θ is α. Note that the units of θ are

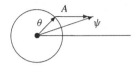

Figure 2.2. The old phase θ is reset by a perturbation of the hand A units to the right, and a new phase ψ results.

radians and those of α are radians/time. Because of this, we refer to α as being a *frequency* and to θ as a phase. The rectangular coordinates of the hand are given by $x = R\cos\theta$ and $y = R\sin\theta$, where R is the length of the hand. Note that this clock goes counterclockwise when $\alpha > 0$. This model of θ is referred to as a *simple clock*.

The simple clock can be thought of as being driven by a motor that moves the hand at a constant rate α, but the human body does not always perceive time in this way. For example, it happens that pacemaker neurons have a base driving frequency plus a modulating frequency that may depend on many things, among them solar time brought into the body through sunlight.

An externally modulated clock is described by the equation

$$\dot{\theta} = \alpha + f(2\pi t/86{,}400),$$

where f has the units of 1/time and has period 2π in its argument. When t increases through one day (86,400 sec), then the modulation f passes through one full cycle. This equation indicates that the hand moves in an irregular way throughout the day. For example, this simple clock is accelerated ($f > 0$) or slowed ($f < 0$) in response to the time of day, perhaps due to sunlight, temperatures, or nutrition.

The ingredients for a modulated simple clock are

a reference time (t),
a center frequency (α), and
a frequency modulation (f).

The mathematical model for a modulated clock is quite similar to the one for a VCO that was described in Chapter 1, and this makes an important connection between model neurons and simple clocks that will enable us to use VCO circuits to model biological rhythms at a level higher than neurons, as shown in Chapter 5.

Modulation of a clock can come from external signals, as the preceding example shows, from feedback of the clock's phase, and from interaction with other clocks. For example, all of these are accounted for by the general model

$$\dot{\theta} = \alpha + F(t, \theta, \psi_1, \ldots, \psi_N),$$

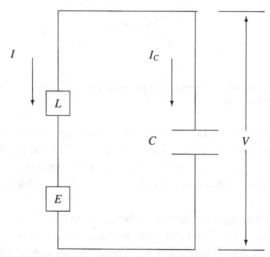

Figure 2.3. LC circuit. L denotes an inductor, E an electromotive force, C a capacitor, and V the voltage across the capacitor. The currents in the circuit are I and I_C.

where $\psi_1, \ldots, \psi_N$ are the phases of other clocks that affect this one. Our approach to modeling networks and uncovering frequency dependence in them in Chapters 5, 6, and 7 is based on VCO circuits, called VCONs, and the models are in terms of phases, essentially like these simple clocks.

2.1.3 Clocks in linear circuits

This and the next subsection show how clocks can be identified in other electrical circuits where they are perhaps not obvious. First, consider an LC circuit as shown in Figure 2.3. If V denotes the voltage across the circuit's capacitor, then

$$C\dot{V} = I_C$$

and

$$L\dot{I} = E - V$$

or, equivalently,

$$LC\ddot{V} + V = E.$$

We can rewrite this equation as a first-order system. First we set $v = V - E$ and $u = -\sqrt{LC}\dot{v}$; then

$$\dot{v} = \frac{-u}{\sqrt{LC}},$$

$$\dot{u} = \frac{v}{\sqrt{LC}}.$$

If we introduce polar coordinates in the LC circuit by setting

$$v = r \cos \theta$$

and

$$u = r \sin \theta,$$

then we get

$$r\dot{r} = v\dot{v} + u\dot{u} = 0,$$

$$r^2 \dot{\theta} = v\dot{u} - u\dot{v} = \frac{r^2}{\sqrt{LC}}.$$

In particular, the equation for the phase variable is (if $r^2(0) \neq 0$)

$$\dot{\theta} = \frac{1}{\sqrt{LC}}.$$

There are several important points illustrated by this simple calculation:

1. Using polar coordinates shifts the problem from physical variables (voltage and current) into phase and amplitude variables. This accomplishes a great simplification – the transformed model can be solved by inspection!
2. The LC model is linear, and, because the frequency ($\dot{\theta}$) does not depend on the amplitude (r), the *isochrons* (i.e., the lines on which phase is constant) are straight lines emanating from the origin.
3. The LC circuit defines a simple clock, and its hand is defined by the vector from the origin to the point $(\cos \theta, \sin \theta)$. The length of the hand can be set arbitrarily long. This circuit defines a simple clock, but it is not a rubber-handed clock since if the hand is stretched, say from length r_0 to length r_1, it stays at the new length ($\dot{r} = 0$).

Isochrons are useful things to keep track of in studies of oscillators. They are determined in the following way. Consider two solutions, one starting at (r_0, θ_0) and the other at (r_1, θ_1). If the corresponding solutions converge in phase, that is, if $\theta_0 - \theta_1$ mod $2\pi \to 0$, then they are said to have started on the same isochron. That is, an isochron is the set of all initial data whose emergent trajectories converge to the same phase variable.

2.1.4 Clocks in nonlinear circuits

Adding a nonlinear element to the simple LC circuit can make a big change! Consider the circuit in Figure 2.4.

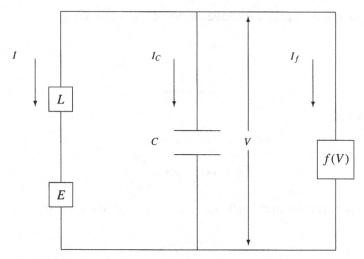

Figure 2.4. A relaxation oscillator. Here E denotes an electromotive force, L is an inductor, C is a capacitor, and $f(V)$ is a circuit element whose IV characteristic is given by $I_f = f(V)$. The currents in the circuit are I, I_C, and I_f.

Balancing currents and voltages using Kirchhoff's laws, we derive the mathematical model

$$0 = I + I_C + I_f,$$
$$V = L\dot{I} + E.$$

Simplifying this using calculations from Chapter 1 gives

$$LC\ddot{V} + Lf'(V)\dot{V} + V = E. \qquad (2.1)$$

This equation is known as van der Pol's equation. He derived it with $f(V) = \lambda(V^3/3 - V)$ to model a triode tube circuit, and he interpreted the results in neurophysiological terms [134].

Some amazing things occur for this model. Let's consider van der Pol's equation where

$$f(V) = \left(\frac{V^3}{3} - \lambda V\right),$$

$E = 0$, and $L = C = 1$. Writing the result as a first-order system gives

$$\dot{V} = -f(V) - U,$$
$$\dot{U} = V,$$

and introducing polar coordinates to this system using

$$V = r \cos \theta,$$

$$U = r \sin \theta$$

we have

$$r\dot{r} = U\dot{U} + V\dot{V}$$

$$= r^2 \left(\lambda \cos^2 \theta - (r^2 \cos^4 \theta)/3 \right),$$

$$r^2\dot{\theta} = V\dot{U} - U\dot{V}$$

$$= r^2 - r^2 \sin \theta \cos \theta \left(\lambda - (r^2 \cos^2 \theta)/3 \right).$$

Dividing by r and r^2 in these equations, respectively, gives

$$\dot{r} = r \left(\lambda \cos^2 \theta - (r^2 \cos^4 \theta)/3 \right)$$

and

$$\dot{\theta} = 1 - \sin \theta \cos \theta \left(\lambda - (r^2 \cos^2 \theta)/3 \right).$$

N. N. Bogoliuboff, a Russian mathematical physicist, pointed out in [8] that if λ is small, then r changes slowly relative to the angle variable θ. Therefore, the equations should be approximately described if the right-hand sides are replaced by their averages with respect to θ. To do this, we replace $\cos^2 \theta$ by

$$\frac{1}{2\pi} \int_0^{2\pi} \cos^2 \phi \, d\phi = \frac{1}{2},$$

$\cos^4 \theta$ by

$$\frac{1}{2\pi} \int_0^{2\pi} \cos^4 \phi \, d\phi = \frac{3}{8},$$

and $\sin \theta \cos \theta$ by 0. The averaged equations are

$$\dot{r} = (r/2)(\lambda - r^2/4)$$

and

$$\dot{\theta} = 1.$$

Straight radial lines are isochrons since two phase variables can converge only if they start at exactly the same value. Therefore, this model is sometimes referred to as being the *radial isochron clock* (RIC) model.

A very important fact to observe at this point is that if λ is a negative number, then solutions (that start with small $r(0)$) converge to zero, but if λ is positive, then $r \to 2\sqrt{\lambda}$ as $t \to \infty$. The value $\lambda = 0$ is a *Hopf bifurcation* value for this system at which the solutions start doing something quite different. A small

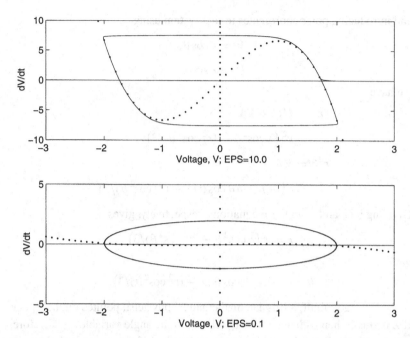

Figure 2.5. Isoclines for van der Pol's equation (dotted lines). Solution for large λ (top). Solution for small λ (bottom).

nonlinear term can cause a major change in the way solutions behave. The smallness of λ appears in the solution in the amplitude of this new periodic solution.

Note that when $\lambda = 0$ the linear model is exactly the LC circuit.

The isoclines and solutions of van der Pol's equation are depicted in Figure 2.5.

2.2 Phase-resetting experiments

The phase-resetting experiments described in the introduction to this chapter are interesting to apply to the rubber-handed clock and to the VCON model. Since such experiments have been carried out for a variety of animals, it is possible to compare predictions made by these theoretical models with what is observed in practice. This is done here first by deriving the predictions for various models and then by comparing them to some observations that are presented in Chapter 5.

The rubber-handed clock gives a nice visualization of phase resetting, but the VCON model is not that simple. Still, we begin with a description of phase resetting of a rubber-handed clock, and then repeat these experiments for the VCON.

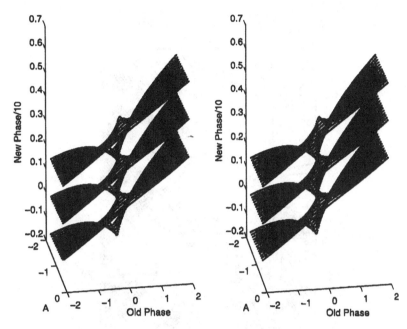

Figure 2.6. Phase-resetting surface.

2.2.1 Phase resetting a rubber-handed clock

Figure 2.2 suggests an interesting experiment. Suppose that a rubber-handed clock is at phase θ and the hand is moved quickly A units to the right, as shown in Figure 2.2, and then released. This shows that the clock is reset at a new phase ψ; in the case shown, the clock is delayed by $\theta - \psi$ radians.

It is possible to calculate ψ in terms of A and θ. In fact, the dot product of the two vectors

$$r_1 = (\cos\theta, \sin\theta) \text{ clock hand}$$

and

$$r_2 = (A + \cos\theta, \sin\theta) \text{ stretched hand}$$

is equal to the product of their amplitudes and the cosine of the angle between them. The result is that

$$1 + A\cos\theta = \cos(\theta - \psi)\sqrt{1 + 2A\cos\theta + A^2}.$$

The diagram in Figure 2.6 is a plot of the new phase ψ as a function of the old phase θ and A. The surface that results is surprisingly complicated since

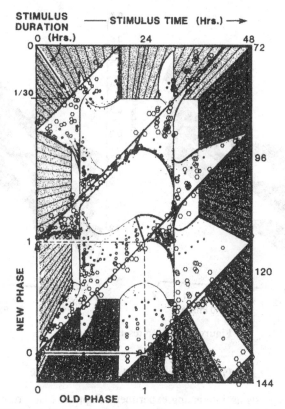

Figure 2.7. Winfree's time crystal for drosophila eclosion. (Reprinted from [142] with permission.)

it has singularities that correspond to the clock's hand being crushed to the center. This method of presenting phase-resetting data was introduced by A. S. Winfree [142].

Evidence of the relevance of a rubber-handed clock to real life phenomena was given by Winfree, who plotted data obtained from fruit fly rhythms. Figure 2.7 is taken from Winfree's book, and it shows data fitting a surface similar to the inverse cosine one derived here.

An interesting point made by Winfree is that the rubber-handed clock has a singularity on its face, namely, at $r = 0$. One impact of this is that if the correct amplitude (in this case, $A = R$) is applied at the correct time (e.g., $\theta = \pi$), then the clock is reset to where its phase is indeterminate. Thus, the timing mechanism is killed. This appears in the time crystal diagram at the vertical lines about which sheets of the surface twist.

Biological timing mechanisms can be quenched by stimulating them at critical times, and the rubber-handed clock's black hole ($r = 0$) suggests one mechanism for this.

Another way in which timing can be destroyed is illustrated by the relaxation oscillator, where a change in the external forcing (E) can change it from having a self-sustained (timing) oscillation to one having a stable equilibrium but no oscillation that can be used for timing. This latter kind of quenching is observed in saddle-node on a limit cycle bifurcations and in the VCON circuit.

2.2.2 Phase resetting of a PLL

We now consider a repetitive VCON (see Section 3.3.5) that is modeled by

$$\dot{x} = \omega + S(V(x)),$$

where $|\omega| > \sup_u |S(u)| \equiv S^*$. $S(u)$ is a sigmoidal function (e.g., tanh(u)) and V is a periodic function. The form of this equation shows that the VCON model is quite similar to a simple clock, except that its progress is modulated by feedback through the controlling voltage. The clock face, or read out, for this clock is described by the circle ($\cos x$, $\sin x$), but the voltage put out by the VCO is also readable – it is $V(x)$, a function that is 2π periodic in x, and it plays the role for the device of the clock's face, too.

A phase-resetting experiment can be performed for this VCON by introducing an inhibitory stimulus $g(t)$: Consider

$$\dot{x} = \omega + S(V(x) - g(t)).$$

For the stimulus g we take a voltage pulse that is described by the tent-shaped function

$$g(t) = \begin{cases} 0 & \text{if } t > T^* \text{ or if } t < t^* \\ (t - t^*)G & \text{if } t^* < t < (t^* + T^*)/2 \\ (T^* - t)G & \text{if } (t^* + T^*)/2 < t < T^*, \end{cases}$$

where $G > 0$ is a constant. The strength of the stimulus is $(T^* - t^*)G$.

The experiment consists of applying the stimulus over some interval of time and then observing the effect. To keep track of things, we keep a reference clock by using this VCON without applying the stimulus to it. Let us denote by x_0 the phase of the unforced VCON, which is determined from the equation

$$\dot{x}_0 = \omega + S(V(x_0)), \quad x_0(0) = x(0).$$

We can now compare $x(t)$ with $x_0(t)$ and thus determine the phase shift due to the stimulus.

Because this equation is identical to the one for $x(t)$ for $t > T^*$, we can define a phase shift due to the stimulus: Let ψ denote a number such that

$$x_0(T^* + \psi) = x(T^*);$$

ψ is either a positive or a negative number, but its amplitude is less than the period of x_0 MOD 2π. The function $y(t) = x_0(t + \psi)$ solves the equation

$$\dot{y} = \omega + S(V(y))$$

for $t > T^*$, and it satisfies the initial condition

$$y(T^*) = x(T^*).$$

x satisfies the same equations, and since this problem has a unique solution, we conclude that $y(t) \equiv x(t)$ for $t > T^*$.

This calculation shows that $x(t)$ looks exactly like $x_0(t)$ for $t > T^*$, except that it is shifted by ψ units in real time. Therefore, we refer to ψ as being the phase shift due to the stimulus $g(t)$.

A surface is plotted in Figure 2.8. A stimulus of strength G is applied at each phase $x(0)$, and the resulting new phase is plotted. Consider the curve traced by the left edge of this surface. Apparently there is little influence of the perturbation except when the period of x_0 is observed to be approximately equal to $4\pi/3$. We take $T^* - t^*$ to be 0.3 and $G = 100$, and we apply g at various times during this period, say at $t = (t^* + T^*)/2$ when $x = x^*$. Then the computed value of ψ is denoted by $\Delta\phi(x^*)$. (See Figure 5.10.)

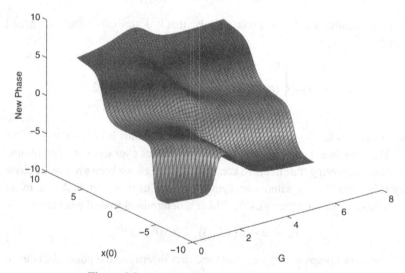

Figure 2.8. Phase-resetting surface for a VCON.

After seeing the response described in Figure 2.8, it is not difficult to predict what effect g will have in general. Since $S(V(x))$ is near $-S^\star$ when $V(x)$ is negative, g will have little influence (recall that g is inhibitory here). On the other hand, when $V(x)$ is positive, then $S(V(x) - g(t))$ is near $S^\star$ for t near $(t^\star + T^\star)/2$, and so g has maximum influence.

Phase-response surfaces similar to the one in Figure 2.8 have been observed in hamster's activity-rest cycles, as we will see in Chapter 5.

2.3 The connection between neurons and simple clocks

Let $V(\theta)$ denote the membrane potential of a nerve cell's hillock region [119] at phase θ. We suppose that V is 2π periodic in θ, so θ increasing by 2π corresponds to one action potential. V is depicted in Figure 3.2. A typical neuron that is firing repetitively might have a period of approximately ten milliseconds, so we write

$$\dot{\theta} = 2\pi \times 10^{-3},$$

and the membrane potential is $V(2\pi \times 10^{-3}t)$. Thus, embedded in the repetitive firing of a neuron is a simple clock. Modulation of the clock's phase is discussed in the next chapter.

As we have seen, a stable limit cycle for a model defines a timer, and progress around the limit cycle is like movement of a hand around the face of a clock. Let this limit cycle be denoted by $p(t)$. Then for any constant ψ the function $p(t + \psi)$ also traces out the limit cycle, and the only difference between $p(t)$ and $p(t + \psi)$ (if $\psi \neq 0$) is that they are like separate hands on the same clock. They differ in phase.

A similar idea applies to solutions starting near, but not on, the limit cycle. Each will approach the limit cycle; some may approach $p(t)$ and each will approach the form $p(t + \psi)$ for some constant ψ that depends on where the solution starts. Therefore, there is a fibration of the space near the stable limit cycle defined by initial points that emanate into solutions that approach $p(t)$, and those that approach $p(t + \psi)$ for each $\psi \neq 0$.

Specifically, to each $\psi \in \overline{0, T}$ where T is the period of p, there is a set of initial points near the limit cycle defining initializing solutions that approach $p(t + \psi)$ as $t \to \infty$. We denote this set by $I(\psi)$. This set is referred to as being the *isochron* determined by phase ψ. How far these isochrons can be extended into the rest of space is a complicated issue and will not be pursued here. However, computer simulation of isochrons easily leads to useful pictures that describe the phase deviation response of the oscillator. See Exercise 6.

2.4 Summary

The first several sections of this chapter introduced ways that clocks can be described mathematically. The idea is to find this kind of description in physical systems and thereby identify them as being timers. This is illustrated by reducing van der Pol's model to a Radial Isochron Clock. Of course, a VCON is presented *a fortiori* as being a modulated simple clock.

This approach suggests how phase-resetting experiments can be interpreted for VCONs and RICs. Real data from phase resetting experiments presented and discussed in Chapter 5 show that phase resetting of the RIC and the VCON agree qualitatively with experiments.

An important feature of van der Pol's model is that it accounts for inputs that force it into nonoscillatory behavior and so destroy its timing ability. For example, forcing the stable oscillation to the unstable equilibrium in the oscillatory case or the forcing potential (E) moving outside the negative resistance region of the escapement stops the oscillation. This feature appears in the VCON through another mechanism – a saddle-node on a limit cycle bifurcation. When the center frequency is forced below a certain threshold (namely, one where a saddle-node bifurcation occurs), then the timing of the VCON is stopped.

The work in the chapter is presented for two reasons. First, the issue of simple clocks and the access they give to important biological phase-resetting experiments is useful in its own right. Second, the use of simple clock models gives us a convenient way to introduce ideas of analysis in the frequency domain without a great deal of technical background. We will use both of these developments later.

Another useful feature of timers that this chapter brings out is that each oscillator in a network can be a timer, and the ensemble then becomes a *multi timer*. That is, if we view the time on the clock faces corresponding to each of the timers in the network, we are observing a very complicated timing device by watching many hands at once. A useful way to think about this is to introduce a phase variable for each oscillator in the network, say we describe these with the variables θ_j for $j = 1, \ldots, N$, and describe the hand on the corresponding clock by the location $(\cos \theta_j, \sin \theta_j)$. Each of these timers progresses at its own pace, and thus we are forced to view all N phases simultaneously; namely, we must consider the vector of phase variables

$$(\theta_1, \ldots, \theta_N).$$

Since each of these phases need only be considered MOD 2π, this vector need only be considered in E^N MOD 2π. This set is called the N-torus, denoted by T^N. (See Exercise 8.)

Our brain comprises a huge number of timers that are effectively monitored simultaneously in a variety of ways. This evidently is an important attribute of our self-awareness.

2.5 Exercises

1. *Phase-resetting of a rubber-handed clock.* Suppose that the rubber-handed clock is at time ψ_0 when its end point is moved A units horizontally and then released, as shown in Figure 2.2. Calculate the change that will result in the observed time if this experiment is performed for

$$\psi_0 = \frac{\pi}{8}, \frac{3\pi}{4}, \frac{5\pi}{4}, \frac{\pi}{2}.$$

2. *Contour lines of a time crystal.* Plot the contour lines of the time crystal derived for the rubber-handed clock on a graph with axes the old phase ψ and A, the stimulation strength. This can be done by printing the values of ψ_1 at each point of a rectangular grid in the A-ψ plane.
3. *Time crystal of a VCON.* Construct a time crystal for the first-order phase-locked loop model. Plot the results and compare them to Winfree's time crystal.
4. *van der Pol's equation in polar coordinates.* Begin with van der Pol's equation, introduce polar coordinates, and obtain the results at the end of Section 2.1.4 for it. This shows that van der Pol's equation can be reduced to a radial isochron clock.
5. *Phase-resetting experiment for the radial isochron clock.* Carry out the phase-resetting experiment described in this chapter for an RIC that is described by the equations

$$\frac{d\rho}{dt} = \varepsilon\rho(1 - \rho),$$

$$\frac{d\psi}{dt} = -1.$$

Apply a pulse to $V (= \rho \cos \psi)$ of varying strength at various phases ψ and plot the results as a time crystal, as shown in Exercise 2.3.
6. Plot the isochrons of van der Pol's oscillator.
7. Define the rotation number for van der Pol's oscillator. Simulate this number for a variety of tunings.
8. Consider two clocks that we monitor simultaneously. Say that they have phases θ_1 and θ_2, respectively. Suppose they are described by the equations

$$\dot{\theta}_1 = \alpha$$

and

$$\dot{\theta}_2 = 1.$$

a. Show that because we are interested in these two variables only through readouts of them through periodic functions, we can consider them as variables MOD 2π and that this is equivalent to considering them on a patch of the plane having side 2π.

b. Use this point of view to show that solutions of these two equations can be depicted as being lines on the surface of a torus. (Hint: For this one can use toroidal coordinates:

$$x = (1 + a \cos \theta_1) \cos \theta_2,$$

$$y = (1 + a \cos \theta_1) \sin \theta_2,$$

$$z = a \sin \theta_1,$$

where a is the radius of the central circle making up the torus.)

c. Show that if α is a rational number, any solution curve of the system will trace out a closed curve on the surface of the torus.

d. Using computer simulation, show that if α is an irrational number (e.g., $\alpha = \sqrt{2}$) then a solution curve appears to be dense on the torus.

e. Using this methodology, plot solar and lunar time clocks.

3

Some mathematical models of neurons

Biological membranes play fundamental roles in many of life's processes. Much of their activity is electrical, and the membrane potential, or voltage across the membrane, is one of the physical states of nerve cells that can be measured *in vivo*. Flows of various ions (charged chemical molecules) through membranes establish electrical currents that cause changes in the membrane potential. These changes are often observed to be pulses of voltage that are called *action potentials*. We are particularly interested here in how action potentials can be described, at what frequency they are generated, and what information they can carry.

Neurophysiology describes the electrical properties of nerve cell membranes. Models of nerves are based on Nernst's equation that determines a cell membrane's potential from the ion concentrations near it. This is described first.

Neurons are described next. A neuron comprises *dendrites* that receive signals, a *cell body* that synthesizes incoming signals and generates new ones, an *axon* that transmits the signals toward other cells, and *synapses* that pass the signals on to other cells. Pulses of voltage received by the dendrites through *chemical* or *electrical synapses* from other cells are combined in the receiving cell, and they might or might not stimulate the formation of a new pulse in the cell. If one is generated, it might propagate along an axon emanating from the cell body and eventually reach a synapse at the terminus of the axon. If the synapse is a chemical one, the pulse will cause release of powerful chemicals, called neurotransmitters, that diffuse across a short gap to interact with a dendrite of another cell; if the synapse is electrical, also called a *gap junction*, the voltage pulse might be communicated directly to the receiving cell.

Many attempts to describe a nerve cell's electrical behavior have been based on electrical circuit analogies. Lapicque [90], van der Pol [134], Katz [83], Hodgkin and Huxley [51], FitzHugh [27], and Perkel et al. [108] are among those who have taken this approach. The Hodgkin–Huxley model provided a major breakthrough: They had formulated a mathematical model that was closely related to experimental data for a patch of axon membrane. Although the model does not accurately describe all aspects of membrane behavior, it has been instrumental in suggesting and understanding a variety of important experiments.

Various models have been formulated that highlight important features of the Hodgkin–Huxley model but are more tractable for mathematical analysis and numerical simulation. These are based on modeling ionic channels, or generalizations of them, and they are of the "fill-and-flush" kind, where a charge builds up and then is released through an escapement within the circuit.

After a brief description of neuron physiology, we develop in sequence ionic channel models, including the Hodgkin–Huxley model, shunting models for multiple channels, single-channel models with escapements (FitzHugh–Nagumo and integrate-and-fire models), and a model in the frequency domain for a single channel with escapement. This leads to formulation of the Voltage Controlled Oscillator Neuron (VCON) model, which is done at the end of this chapter.

Modeling neurophysiological systems is quite complicated. For example, one can focus at the molecular level and follow the flow of ionized molecules through membrane pores and, as a result, uncover behavior of membranes on a small scale as Hodgkin and Huxley did. One can also, however, model large systems of neurons by capturing certain dominant features and lumping complicated mechanisms into tractable ones. Mathematical modeling is used at both of these levels. Electrical circuits describing ion flows are stated and studied using mathematical terms, but integrative systems are also described in mathematical terms. Moreover, one passes from a molecular channel level to an integrative system level using mathematical methods. It is especially useful to study such networks in the *frequency domain* since that uncovers how information can be carried by timing of cell signals. The work in this chapter lays a foundation for moving from the channel level of modeling to modeling large systems, which is carried out in Chapters 5, 6, and 7.

3.1 Neurophysiology

Neurophysiology is one of the shining stars in all of science. Amazing experimental techniques have been developed and used to uncover the existence of neurons and how they work individually and in networks. The most successful methods to date have been at the molecular level, and the future will see passage of knowledge about this level on to higher levels of organization. Mathematical modeling will play a central role in these developments [50].

The work here is based on several experimentally determined details:

- Ionic currents through membranes are known to be responsible for action potentials, and they propagate along membranes [51].
- There is a relation between the timing (frequencies and phase deviations) of inputs to membranes and the timing of their firing [43].
- Neurons communicate through excitatory and inhibitory connections [88].
- There are electrical synapses, or gap junctions, and there are chemical synapses [81].

We will use these facts in our modeling here and in later sections.

3.1.1 Nernst's equation

The physical dimensions of mass (M), length (L), charge (Q), temperature $(\mathbf{T})$, and time (T) are used in describing membrane behavior. *Dimensional analysis* is useful to keep careful track of which terms balance in models. Underlying dimensional analysis is the fact that in an equation both sides must have the same physical units. It is a simple, but very important fact to keep in mind. To use dimensional analysis one must know the dimensions of several important physical quantities. For example, volume is given in units of length cubed, so we write $[volume] = L^3$. Using these ideas, we have the following:

$$[volume] = L^3,$$
$$[concentration] = [mass/volume] = M/L^3,$$
$$[flow] = [volume/time] = L^3/T,$$
$$[flux] = [mass/area/time] = M/L^2T,$$
$$[force] = [mass \times acceleration] = ML/T^2,$$
$$[pressure] = [force/area] = M/LT^2.$$

For more details on how physical models of cell dynamics are described, see [56].

An electrical potential is established across a cell's membrane by having different concentrations of electrically charged chemical species inside and outside. For example, if a semipermeable membrane separates two regions of space that have concentrations of ions, say C_1 and C_2 inside and outside, respectively, then a short calculation [56] shows that

$$-qE = k\mathbf{T}\log\frac{C_2}{C_1}.$$

Solving for E, we have Nernst's equation

$$E = \frac{k\mathbf{T}}{q} \log \frac{C_1}{C_2}.$$

Here $\mathbf{T}$ denotes temperature in degrees Kelvin, k is Boltzmann's constant, and q is the charge on each ionized molecule. This formula shows how to compute the membrane potential once the ion concentrations inside and outside are known.

3.1.2 Cell membrane potentials

The most important ions to the neuron membrane potential are sodium (Na^+) and potassium (K^+). Each ionic species has associated with it a membrane potential that is maintained by pumps in the membrane that force Na^+ out and K^+ into the cell. At equilibrium, Nernst's equation shows what the potential for each of these ionic species is: For sodium

$$E_{Na} = \frac{k\mathbf{T}}{q} \log \frac{C_o^{Na}}{C_i^{Na}} = 55 \text{ mV}$$

and for potassium

$$E_K = \frac{k\mathbf{T}}{q} \log \frac{C_o^{K}}{C_i^{K}} = -75 \text{ mV}.$$

These are referred to as being the sodium and potassium resting potential, respectively, and they are measured from the outside to the inside.

3.1.3 Action potentials

A nerve cell is depicted in Figure 3.1. The main parts of interest here are the *dendrites*, which receive signals from impinging axons; the *cell body* that sums inputs and can create electrical activity, usually in the form of a voltage pulse called an *action potential*; the *axon*, which carries the action potential away from the cell body to a synapse; and the *synapse*, which causes electrical or chemical signals to be released outside the cell in response to the arrival of an action potential. *Chemical synapses* release chemical signals in the form of molecules called *neurotransmitters*. These diffuse across the synaptic gap to a dendrite, or to the cell body, and cause an electrical potential imbalance to be created across the postsynaptic membrane. Chemical synapses are not well understood since, for example, contractile proteins in the postsynaptic membrane can cause the gap size to change [82], but much is known about them. *Electrical synapses*, or gap junctions, transmit electrical signals directly from the sending cell to

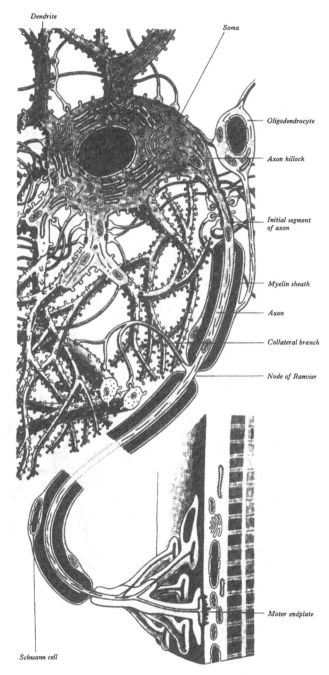

Figure 3.1. Schematic drawing of a motor neuron indicating the dendrites, the cell body, the axon hillock, the axon and its terminus. (Redrawn from Figure 8.28 in [40] with permission.)

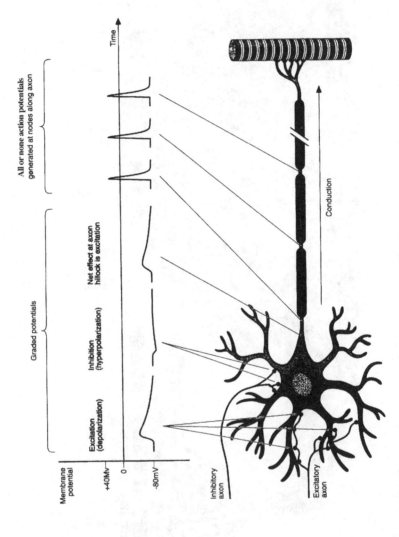

Figure 3.2. Types of change in membrane potential that can be recorded at the sites indicated. (Redrawn from Figure 8.6 in [40] with permission.)

the receiving cell. Eventually, these can either drive the cell body toward firing (i.e., toward creating an action potential) or inhibit it from firing.

Neuron membranes at rest are electrically neutral. When excited, Na^+ channels open rapidly and the membrane potential increases toward E_{Na}. The K^+ channels open more slowly, but eventually they return the membrane potential toward E_K. These close, and other cell mechanisms bring the cell membrane back to neutral. This is depicted in Figure 3.2. The mechanisms that control these channels remain unknown, although many models of membrane potential and currents have been derived and used effectively to suggest experiments and interpret data.

3.1.4 Synapses

Presynaptic and postsynaptic cell membranes can be in direct contact with each other and pass electrical information between them. Changes in the membrane potential might be caused by direct passage of ions, for example Ca^{++}, from one cell to the other, forced electrostatically by an arriving action potential. However, exact mathematical modeling of gap junctions has not yet been done.

Chemical synapses are better understood. Action potentials arriving at a synapse (called *presynaptic potentials* here), cause vesicles containing chemical neurotransmitters to migrate to the synapse membrane and release their contents into the synaptic gap. This is depicted in Figure 3.3.

Over thirty neurotransmitters are now known, and others and the roles they play will probably soon be discovered. For example, acetylcholine (ACh) is an excitatory neurotransmitter and γ-aminobutyric acid (GABA) is an inhibitory one [88].

Neurotransmitters diffuse across a synaptic gap, though some are lost from the gap due to diffusion out of it. The molecules that arrive at the postsynaptic membrane interact with it to modify its membrane potential. Let $c(t)$ denote the concentration of neurotransmitter in the gap at time t. It increases in response to the arrival of action potentials and decreases because of chemical binding with the postsynaptic membrane and because of diffusion out of the gap. The chemical kinetics are modeled by the equation

$$\dot{c} = -k_{\text{diff}}c - k_{\text{post}}c + S,$$

where the rate constants k_{diff} and k_{post} describe diffusion from the gap and binding with the postsynaptic membrane, respectively. Here S denotes a source of neurotransmitter due to the vesicle contraction caused by arriving action potentials.

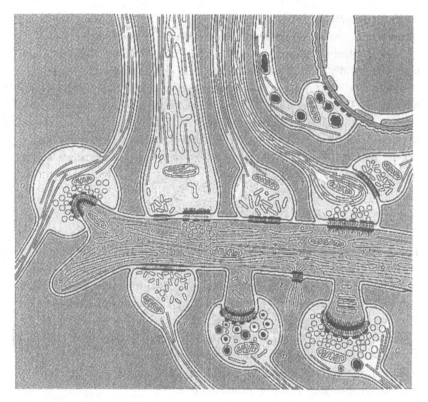

Figure 3.3. Scheme of various chemical structures grouped around a dendrite. (Redrawn from Figure 8.35 in [40] with permission.)

The postsynaptic interaction remains largely unknown, but we make an assumption that the postsynaptic membrane potential is augmented or diminished by excitatory or inhibitory neurotransmitters, within limits above and below which there is no further effect. That is, the response is clipped by physiological limits of the membrane's response.

3.2 Ionic channel models

Hodgkin and Huxley [51] described how ionic currents passing through the membrane of a neuron can create spikes in the membrane's voltage that propagate along the membrane and so carry information. As part of their work, they proposed an electrical circuit analogue model of ionic channels through a nerve cell membrane. We study this in several steps, beginning with a circuit analogue of a single channel.

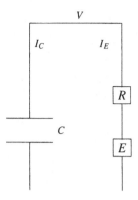

Figure 3.4. Circuit analogue of a single ion channel.

3.2.1 A single-ion channel

First, we consider ion flow through a single channel. Our model is the circuit described in Figure 3.4.

This circuit accounts for capacitance of the membrane since when it is impermeable to the ion, it acts as a capacitor storing charge. When the channel is open there is a resistance to the flow of ions, reflected by the resistor R (having conductance $g = 1/R$), but the potential equilibrates to the ion's resting potential, here denoted by E. The resting potential is maintained by ion pumps in the membrane and other cell mechanisms.

The mathematical model for this circuit follows from Kirchhoff's law

$$I_C + I_E = 0,$$

reflecting that charge is conserved in any circuit; the equation

$$I_C = C\dot{V},$$

which describes how a capacitor works; and Ohm's law

$$RI_E = (V - E),$$

which relates the current through a resistor to the voltage across it. It is more convenient to think in terms of the conductance of the resistor rather than its resistance since the channel being closed is described by $R = \infty$. Therefore, we write $g = 1/R$. When $g = 0$ the channel is closed, and when $g > 0$ it is open.

As a result, we have

$$C\dot{V} = g(E - V),$$

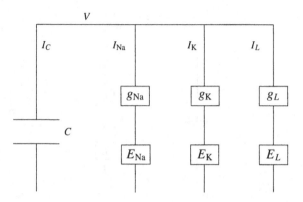

Figure 3.5. A circuit with several channels: V denotes the membrane potential, C is the membrane capacitance, and the three ionic channels are modeled by a resting potential (E_{Na}, etc.) and a conductance (g_{Na}, etc.) for sodium (Na^+), potassium (K^+), and leakage, respectively. The resistances R_{Na}, etc. are reciprocals of the conductances $g_{Na}(V) = 1/R_{Na}$, etc.

from which we see that

$$V(t) = E + \exp\left(-\frac{gt}{C}\right)(V(0) - E).$$

When the channel is open, $V(t) \rightarrow E$ at a rate proportional to the channel's conductance.

3.2.2 Sodium, potassium, and leakage channels

Next, we consider a membrane that has several channels through which various ions flow, as illustrated in Figure 3.5.

The mathematical model for this circuit follows directly from Kirchhoff's laws. As before, we have

$$C\dot{V} + g_{Na}(V - E_{Na}) + g_K(V - E_K) + g_L(V - E_L) = 0, \qquad (3.1)$$

where

g_{Na} and E_{Na} are the sodium conductance and resting potential, respectively;

g_K and E_K are the potassium conductance and resting potential, respectively;

g_L and E_L are the conductance and resting potential, respectively, for ions passing through the membrane by other mechanisms, such as ion pumps, facilitated diffusion, passive diffusion, and leakage; and

$C\dot{V}$ has the units $\mu A/cm^2$, which is in the units of current per unit area and reflects the ionic current through a patch.

3.2.3 The Hodgkin–Huxley phenomenological variables

The circuit in Figure 3.5 accounts for migration of the primary ions involved in action potential generation (sodium and potassium).

The sequence of events in this process comprises:

1. Through external or internal stimulation, the sodium channel opens, so $g_{Na} > 0$. This conductance is relatively large.
2. The potassium channel opens and the sodium channel closes, so $g_{Na} = 0$ and $g_K > 0$.
3. The potassium channel closes and ion leakage resets the membrane to its resting potential E_L.

The circuit described by Equation (3.1) is the one used by Hodgkin and Huxley in their studies. The tricky part in their work on this circuit involves describing how the sodium and potassium conductances change with changing values of V. In their experimental studies Hodgkin and Huxley introduced three phenomenological variables, m, n, and h. They use the following definitions and data:

$$g_{Na} = 120 \, m^3 h,$$

$$g_K = 36 \, n^4,$$

$$g_L = 0.3,$$

$$\dot{m} = \alpha_m(1 - m) + \beta_m m,$$

$$\dot{h} = \alpha_h(1 - h) + \beta_h h,$$

$$\dot{n} = \alpha_n(1 - n) + \beta_n n,$$

$$\alpha_m = 0.1(25 - V)/\left[\exp(2.5 - 0.1V) - 1\right],$$

$$\beta_m = 4\exp(-V/18),$$

$$\alpha_h = 0.07\exp(-V/20),$$

$$\beta_h = 1\Big/\left\{\exp\left[(30 - V)/10\right] + 1\right\},$$

$$\alpha_n = 0.01(10 - V)/\left[\exp(1.0 - 0.1V) - 1\right],$$

$$\beta_n = 0.125\exp(-V/80).$$

These numerical values were used to describe the sodium activation (m), inactivation (h), and potassium activation (n) variables when the temperature is 6.4°C [51, 52].

These equations can be solved using computers, and their solutions can be plotted in various interesting ways as explored in the exercises.

One interesting feature of this model is refractoriness: After an action potential has been generated, there is a period of time during which new stimulation will generally not excite another action potential. This occurs during most of the interval when $g_K > 0$.

These equations describe the evolution of an action potential once the system has been excited by an injected current. They were derived to describe a patch of a squid's giant axon, not an entire neuron. Modeling of an entire neuron will require modeling the dendritic field [112, 113], the cell body, the hillock region, the axonal field, chemical transport in the axons, the filling and flushing of vesicles containing neurotransmitters in axonal termini, and chemical and electrical synapses. In addition, descriptions are needed for a nerve cell's metabolism; its interaction at quantum mechanical levels with ions passing inside, into, and out of its membrane; its behavior in the ever changing chemical bath in which neurons live; and how all these influences control a cell's life.

3.2.4 General shunting multiple-channel models

Any number of electrical channels can be modeled using an equation of the form

$$C\frac{dV}{dt} = \sum_{j=1}^{m} g_j(M)\big(E_j^\infty - V\big),$$

where m = number of channels, E_j^∞ is the membrane's resting potential for ions of type j, and $g_j(M)$ describes the conductance of the corresponding ionic channel as a function of a collection of physical or phenomenological variables, $M = (M_1, \ldots, M_m)$.

If one of the conductances is high and the others are low, say $g_k = 1$ and $g_j = 0$ for $j \neq k$, then we see that $V \to E_k^\infty$ as $t \to \infty$. This is referred to as a *channel shunting* model since it provides alternate pathways for current that are gated by the variables in M. The model is quite similar to a bank of low-pass filters whose time constants and electromotive forces adapt to the other variables.

This approach to membrane modeling facilitates accounting for the switching on and off of external excitation and inhibition. To do this we set the appropriate conductances to either positive values or zero. The conductances are assumed to vary, as in the Hodgkin–Huxley model, but now rather than depending on internal phenomenological variables, like m, h, and n, the conductances g_m can be changed directly. For example, an excitatory stimulus to the cell is accounted for by switching g_1 to positive values. The work of Carpenter and Grossberg [14] and others discuss a variety of ways in which this can happen.

It now becomes difficult to associate an electrical circuit with the model, and for this reason, we do not discuss it in later sections. Moreover, we wish to study timing aspects of networks, and this is not directly accessible in shunting models. Still, this approach gives rise to interesting dynamical predictions that can be tested experimentally [42].

The conductances $\{g_j\}$ might be taken to be proportional to nearby concentrations of excitatory and inhibitory neurotransmitters, respectively. These concentrations change in response to external stimulation from impinging synapses and from self-stimulation.

Synapses can deplete their supply of neurotransmitter stored in vesicles. This process can be modeled by taking g_1 to be related to the neurotransmitter concentration in the vesicles, say z. The kinetics of z might be described by the equation

$$\dot{z} = \alpha(S - z) - \beta I z,$$

where I is the stimulating current. In the absence of stimulation ($I = 0$), $z \to S$ at a rate α that is probably slow compared to the maximum depletion rate. Here S denotes a constant source of neurotransmitter, supplied through the cell's metabolism, and β is the rate at which neurotransmitter is released per unit stimulation. When I is switched on, z decreases toward $\alpha S/(\alpha + \beta I)$. The larger the stimulus, the more transmitter is released and so the higher is the excitation conductance. However, then z approaches a low level at rate βI, eventually shutting down the excitation channel. This is referred to as *synapse gating*, and the phenomenon is essentially one of synapse fatigue.

Cohen and Grossberg [16] study general shunting models of the form

$$\dot{a} = \alpha(a)\big(\beta(a) + \gamma(a)Ca\big),$$

where the variables $a_1, \ldots, a_N$ describe "activity" of the channels $1, \ldots, N$. We will return to this model when we consider networks.

3.2.5 A single-channel circuit with an escapement

A variety of electrical circuit models have been derived in efforts to capture voltage-dependent switching aspects of channels. Figure 3.6 depicts a typical one.

Here L denotes an inductor in the circuit, g a resistor having resistance $R = 1/g$, and $f = f(V, t)$ is a switching element in the circuit. The mathematical

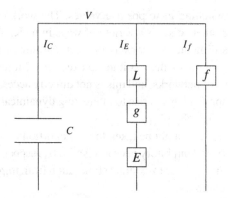

Figure 3.6. Single channel circuit with escapement. The circuit elements are a battery (E), a resistor (R, with IV characteristic $V_R = RI$ (Ohm's Law)), an inductor (L, with IV characteristic $V_L = L\dot{i}$), a capacitor (C, with IV characteristic $I_C = C\dot{V}$), and a nonlinear circuit element (f, with IV characteristic $I_f = f(V, t)$, where f is a nonlinear function).

model of this circuit is

$$C\dot{V} + f(V, t) + I_E = 0,$$

$$L\dot{I}_E + RI_E + E = V,$$

where $I_f = f(V, t)$ is the current-voltage relation for the switching channel. If t does not appear explicitly in the system, it can be studied using phase-plane methods, and it falls into the general category of forced van der Pol equations.

3.2.5.1 The FitzHugh–Nagumo circuit

Switching-circuit elements are described in terms of their IV-characteristics, such as Ohm's law for resistors, or in terms of their input–output relations. Various useful circuit elements have more complicated characteristics. For example, a *tunnel diode* is an electronic device. It has an N-shaped characteristic, $I = f(V)$. The voltages V for which the slope of the characteristic is positive are resistance regions, as in Ohm's law, and the interval where the slope is negative is the negative resistance region. Negative resistance results from curious electronic properties of the material making up the diode and its design.

Tunnel diodes are not popular with circuit designers since they are unstable [68]. However, similar circuits can be constructed using operational amplifiers [57, 85]. This circuit is called the FitzHugh–Nagumo circuit when f is a cubic function of V describing the IV-characteristic of a tunnel diode [27]. The model is for a single channel with a nonlinear switching circuit element that is created by a tunnel diode.

This mathematical model is quite interesting. It is related to earlier circuits, including the van der Pol and Hodgkin–Huxley models, and consequently has been widely studied.

The FitzHugh–Nagumo circuit is typical of various "fill-and-flush" circuits. Applying Kirchhoff's laws to the circuit in Figure 3.6, we get the mathematical model

$$L\dot{I} = -E + V - RI,$$
$$C\dot{V} = -I - f(V),$$

where $I = I_E$.

The first equation comes from balancing the potential drops across all of the circuit elements in the center channel with V. The second equation results from balancing the currents in the circuit.

Such a system of nonlinear differential equations proves challenging to study, but because there are only two differential equations, compared with four in the Hodgkin–Huxley model or more in the shunting models, many more mathematical methods are available for studying it. The standard analytic approach proceeds in two steps:

1. Find the *isoclines*. That is, find the curves in the V-I plane on which $\dot{I} = 0$ and those on which $\dot{V} = 0$. Intersections of these lines identify the *static states* or *equilibria* of the circuit.

2. Determine the linear stability of the static states. This step involves some algebra. First, the nonlinear term is approximated by one linear term near the static state using Taylor's expansion of f. The resulting linear system is then solved by the Laplace transform method (see Appendix A). This shows whether or not solutions starting near the equilibrium approach it as t increases. If they approach it, then the static state is called *stable*; if they oscillate around it, it is called *oscillatory*; and if any solutions of the linear approximation diverge from it, the state is called *unstable*.

When the (linear) stability of a state changes, we say that a *bifurcation* has occurred. Various things can happen in such an instance depending on the dynamics of the system. Mathematicians have studied bifurcations in a variety of settings ranging from quite practical (Newton's polygons), to theoretical (the Implicit Function Theorem), to quite abstract (the theory of singularities). The FitzHugh–Nagumo model illustrates two kinds of bifurcations: Hopf bifurcation and saddle-node bifurcation.

The isocline $\dot{I} = 0$ is determined from the straight line

$$I = g(V - E).$$

This equation describes a straight line with slope g crossing the $I = 0$ axis at $V = E$. Let $-s$ denote the most negative slope of the tunnel diode characteristic curve. We will illustrate two interesting cases here: $-g < -s$ and $-s < -g$.

In the first case, we see that for whatever choice of E, the static potential is uniquely determined. However, it is not stable for values where the isocline crosses the negative resistance region of the tunnel diode! The system oscillates in this case. Therefore, as E changes from high values, the system moves from having a stable rest point (static potential) to one having an unstable static potential and a stable oscillation. This oscillation eventually disappears until a stable equilibrium is again reached. When the crossing is near the appearance of the stable oscillation, the system is said to be *excitable* since a perturbation of the the solution from the equilibrium can initialize a solution that executes a large excursion before returning to the equilibrium.

In the second case, an interval of E values exists over which there are three static potentials. In that case, the interior one is unstable and the other two are stable. This bistability of the system indicates the potential for hysteresis in its solutions. In particular, as E decreases from high values there is one stable solution that will persist until it disappears through a saddle-node bifurcation. The system then jumps to the one remaining static state. As E is increased from this regime, the system stays on the lower static state until it eventually disappears through a saddle-node bifurcation, and the system jumps to the upper static state.

The response of this circuit to external forcing is quite complicated, and it will not be pursued here (see [28]). However, a simplification of this model does enable us to study the effects of forcing on the circuit. This is the *integrate-and-fire* model that we study next.

3.2.6 *Forced single channel–escapement circuit: integrate-and-fire model*

A simple appearing model that finds its roots in the early part of this century [90, 6, 132] remains quite useful today. This is the integrate-and-fire model. The analogous circuit is shown in Figure 3.7.

The circuit consists of a low-pass filter in parallel with an electrical switch (e.g., a field effect transistor). When v reaches a threshold level (v_T), the switch closes, thereby resetting the membrane potential to 0, the resting potential. This action mimics a neuron firing.

We consider the circuit model described by the equations

$$\tau \dot{v} + v = E_0 + E_m \cos(\omega t + \phi),$$

$$\text{if} \quad v(t^-) = v_T \quad \text{then} \quad v(t) = 0.$$

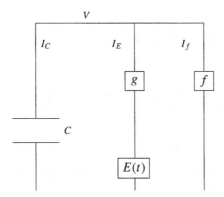

Figure 3.7. Integrate-and-fire circuit. Here the nonlinear element is a simple switch that closes when the voltage V reaches a threshold level v_T. $E(t)$ includes the external forcing on the circuit.

Here

E_0 is the applied constant stimulus (direct current bias),
E_m is the applied alternating stimulus amplitude (alternating current),
$\tau = C/g$ is the filter's time constant,
ω is the forcing frequency, and
ϕ is the phase deviation of the applied alternating current.

We can rescale variables in this model by setting

$$\tilde{t} = \omega t + \phi, \quad \tilde{v} = \frac{v}{v_T}, \quad \sigma = \frac{1}{\omega \tau}, \quad E = \frac{E_0}{\omega \tau v_T}, \quad B = \frac{E_m}{E_0}.$$

Then the model becomes

$$\tilde{v}' = -\sigma \tilde{v} + E(1 + B \cos \tilde{t}),$$

$$\text{if} \quad \tilde{v}(\tilde{t}^-) = 1 \quad \text{then} \quad \tilde{v}(\tilde{t}) = 0,$$

where $\tilde{v}' = d\tilde{v}/d\tilde{t}$. At this point, we drop the tilde notation and thus consider the system

$$\dot{v} = -\sigma v + a_0 + a_1 \cos t,$$

$$\text{if} \quad v(t^+) = 1 \quad \text{then} \quad v(t) = 0,$$

where $a_0 = E$ and $a_1 = EB$.

Integrating this equation from a firing, say at t_n, to t gives

$$v(t)e^{\sigma t} = \frac{a_0}{\sigma}(e^{\sigma t} - e^{\sigma t_n}) + \frac{a_1}{\sqrt{1 + \sigma^2}}(G(t) - G(t_n)),$$

where

$$G(t) = e^{\sigma t}(\sin t + \sigma \cos t)/\sqrt{1 + \sigma^2}.$$

If we choose a number ψ such that

$$\sin \psi = \frac{\sigma}{\sqrt{1 + \sigma^2}}, \qquad \cos \psi = \frac{1}{\sqrt{1 + \sigma^2}},$$

then

$$G(t) = e^{\sigma t} \sin (t + \psi).$$

If $v(t_n) = 0$ and the next firing occurs at t_{n+1} when $v(t_{n+1}) = 1$, then we have

$$e^{\sigma t_{n+1}} = \frac{a_0}{\sigma}(e^{\sigma t_{n+1}} - e^{\sigma t_n}) + \frac{a_1}{\sqrt{1 + \sigma^2}}\big(G(t_{n+1}) - G(t_n)\big).$$

Rearranging terms in this expression gives

$$\frac{a_0}{\sigma}e^{\sigma t_{n+1}} + \frac{a_1}{\sqrt{1 + \sigma^2}}G(t_{n+1}) - e^{\sigma t_{n+1}} = \frac{a_0}{\sigma}e^{\sigma t_n} + \frac{a_1}{\sqrt{1 + \sigma^2}}G(t_n).$$

Multiplying both sides by σ and using the definition of ψ gives

$$e^{\sigma t_{n+1}}\big(a_0 + a_1 \sin \psi \sin (t_{n+1} + \psi) - \sigma\big) = e^{\sigma t_n}\big(a_0 + a_1 \sin \psi \sin (t_n + \psi)\big).$$

As a result, if we define

$$F(t) = e^{\sigma t}\big(a_0 + a_1 \sin \psi \sin (t + \psi) - \sigma\big)$$

then at the next firing, which is denoted by t_{n+1}, we have

$$F(t_{n+1}) = F(t_n) + \sigma e^{\sigma t_n}.$$

Note that the derivative of the function F is given by

$$F'(t) = \sigma e^{\sigma t}(a_0 - \sigma + a_1 \cos t).$$

The relation between the nth firing (i.e., the nth time the voltage hits the threshold), denoted by t_n, and the next at t_{n+1} can be studied. We focus attention on the mapping

$$t_n \to t_{n+1}.$$

The sequence $\{t_n\}$ is referred to as being the *firing phase sequence*. We have

$$F(t_{n+1}) = F(t_n) + \sigma e^{\sigma t_n},$$

which defines the next firing time implicitly from the present one. We suppose here that

$$a_0 - \sigma > a_1 \geq 0.$$

With this assumption, F is a monotonically increasing function mapping the interval $[0, \infty)$ onto the interval $[F(0), \infty)$. Moreover, the values of $F(t) + \sigma e^{\sigma t}$ lie in the range of F. Therefore, the inverse function F^{-1} is defined, and we can write

$$t_{n+1} = f(t_n),$$

where $f(t) = F^{-1}(F(t) + \sigma e^{\sigma t})$. This approach was developed and studied in [85, 114], and a similar model having a constant forcing but oscillatory threshold was studied in [38].

Because the forcing is periodic in our model, the mapping f has an interesting property. Note that

$$
\begin{aligned}
F[f(t)] &= F(t) + \sigma \exp(\sigma t) \\
&= e^{-2\pi\sigma} \left\{ F(t + 2\pi) + \sigma \exp[\sigma(t + 2\pi)] \right\} \\
&= e^{-2\pi\sigma} F[f(t + 2\pi)]. \\
&= F[f(t + 2\pi) - 2\pi].
\end{aligned}
$$

Therefore,

$$f(t + 2\pi) \equiv f(t) + 2\pi$$

for all t. This is referred to as being the *circle mapping property*. In fact, if we define

$$\theta_n = t_n \text{ MOD } 2\pi$$

then for some integers M and N, $t_n = \theta_n + 2\pi N$ and $t_{n+1} = \theta_{n+1} + 2\pi M$, and we have that

$$
\begin{aligned}
\theta_{n+1} &= t_{n+1} - 2\pi M \\
&= f(t_n) - 2\pi M \\
&= f(\theta_n + 2\pi N) - 2\pi M \\
&= f(\theta_n) + 2\pi(N - M) \\
&= g(\theta_n),
\end{aligned}
$$

where $g(\theta) = f(\theta) \text{ MOD } 2\pi$. This follows from the circle mapping property, and it enables us to reduce consideration of the firing times determined by f to their representations on a circle described by g.

The rotation number is defined for this iteration by the formula

$$\rho = \lim_{n \to \infty} \frac{f^{[n]}(\theta_0)}{2\pi n},$$

where θ_0 is any initial point, $f^{[0]} = f$, and for $n \geq 1$, $f^{[n+1]}(x) = f(f^{[n]}(x))$. This number was studied by Denjoy [22], and he showed that if ρ is a rational number, then the sequence of firing phases is periodic, but if the rotation number is irrational, then the sequence is aperiodic, or ergodic. Moreover, he showed that the rotation number depends continuously on the data of the problem, in this case, on τ, E_0, E_m, ω, and ϕ.

Some further analysis is possible to describe phase locking and ergodic behavior in the integrate-and-fire model, but usually computer simulations are used at this point. The rotation number is difficult to calculate using analytic methods, but it is quite useful for visualizing computer simulations. We use it repeatedly in the rest of this book.

Computer simulations of rotation numbers have been used to study repetitive firing behavior in biological systems employing the integrate-and-fire model, which can be analyzed as described above, and to study more general models.

Our analysis of the integrate-and-fire model made use of a mapping of the unit circle into itself. The variables used in that case were phase (or angle) variables, and our analysis of the problem describes how it can be converted from physical variables to the frequency domain. This enables us to use the well-developed theory of rotation numbers, which is a frequency-domain methodology.

3.2.7 A channel-and-escapement model in the frequency domain

Channel-and-escapement models can be studied in the frequency domain. Consider the circuit in Figure 3.6 with $L = 0$ and the switching element now changing with time ($f(V, t)$), say

$$C\dot{V} + f(V, t) + g(V - E) = 0.$$

In this equation the nonlinearity is a cubic, similar to a tunnel diode, but the coefficients in it depend explicitly on t, reflecting input to the circuit that comes through the switching mechanism.

For example, consider the (normalized) equation

$$\tau\dot{V} = -\left(V^3 - A(\theta)V + B(\theta)\right),$$

where τ is a time constant and A and B are 2π-periodic functions of a new variable θ. This variable describes how the data evolve with time. θ might depend on exogenous time t, and it might proceed faster at some times than others. For example, in some places, say $V < 0$, the cell is refractory, and input has little impact on the circuit. To describe these possibilities, we write

$$\dot{\theta} = \mathcal{F}(\theta, t), \tag{3.2}$$

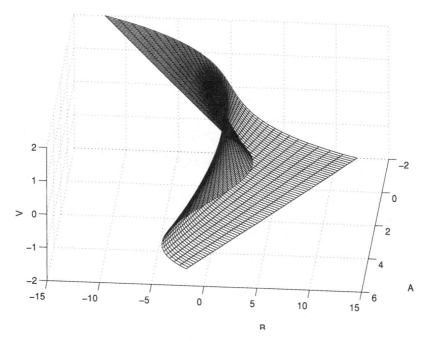

Figure 3.8. The cusp surface.

where $\mathcal{F}$ describes the way in which θ changes with t. We can gain some insight to the behavior of solutions to this equation by recalling some facts about the cusp surface that is described by the equation

$$B = Av - v^3.$$

The surface is depicted in Figure 3.8.

When $A = r \cos \theta$ and $B = r \sin \theta$ the variables (A, B) trace out a circle about the origin. The corresponding point on the cusp surface is uniquely determined outside the overlap region, but it is triply defined inside of it. The solution of the differential equation

$$\tau \dot{V} = -(V^3 - AV + B)$$

has the cusp surface as its equilibrium set. If $\tau \ll 1$, as the point (A, B) moves in a circle about the origin, the solution tracks the circle moving along the lower branch, around behind it, up to the upper branch, and then falling off the fold to return to the lower branch. This can be seen in another way by looking straight down on the surface, as shown in Figure 3.9.

If the parameter $\tau \ll 1$, then the singular perturbation solution of this system will be $V = V^*(\theta)$, where V^* is a fixed wave form resembling a

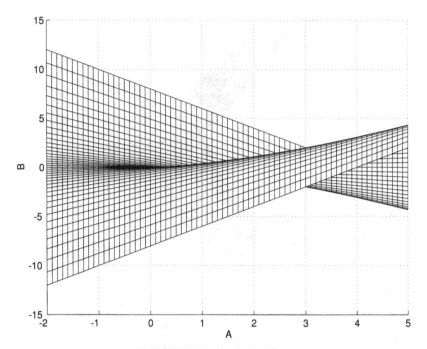

Figure 3.9. Cusp variables A, B.

lock washer [58]. As a result, the circuit's dynamics are actually described by Equation (3.2).

In particular, we have derived here a phase-variable rendition of a one-channel circuit with escapement model, similar to the FitzHugh–Nagumo and integrate-and-fire circuits. This shifts the focus from physical variables of voltages and currents to the frequency domain [68] where a number of further calculations are possible, including analyzing phase-locking phenomena that are so important in neural networks.

These topics are studied later. We next introduce a more general model in the frequency domain.

3.3 Neuron modeling in the frequency domain

As shown in the preface to this book, models in phase variables arise naturally when general systems are near saddle-node on a limit cycle bifurcations. In that case, a model resembling a frequency-domain model emerges as being the canonical model for a saddle-node bifurcation (see Appendix B). In the previous section, we saw that channel models can be described in the frequency domain, and we saw earlier that circuits important to signal processing were designed in the frequency domain. In this section we take the next step in our study of signal processing attributes of networks: We derive a neuron circuit analogue in the frequency domain: the VCON.

The VCON model was developed in 1979 [57] [see 65] to model a part of a neuron, namely the hillock region of an axon where action potentials are generated [119]. The hillock region was visualized as being a VCO, and one ignored the dendritic field, the cell body, and the transmission time on an axon. VCONs could be coupled by constructing analogues of chemical and electrical synapses. The VCON model is derived and studied in this section.

3.3.1 The cell body's action potential trigger region

Neurons studied here operate in either a repetitive firing mode (pacemakers) or an excitable mode, similar to a saddle-node on a limit cycle bifurcation. We have seen similar behavior in VCOs. Recall that a VCO is modeled by the equation

$$\dot{\theta} = \omega + V_c,$$

where ω is the center frequency, V_c is the controlling voltage, and θ is the phase of the VCO's output voltage.

The VCO output, say $V(\theta)$, can be either a square wave, a triangular wave, or sinusoidal. However, virtually any desired profile of V can be constructed by passing the VCO output through a wave form synthesizer. In particular, the action potential form can be synthesized, although we see in Figure 3.10 that sinusoidal VCO output bears a striking resemblance to an action potential. The wave form is not particularly important; it is the phase variable and how it changes as a function of t that makes it possible to trace out an action potential using a sinusoidal wave form for V. Shown in Figure 3.10 is the wave form output $\cos(\theta + \pi/4)$ when θ is determined from the equation

$$\dot{\theta} = 1 + \cos\theta.$$

In the work that follows, we take V to be any continuously differentiable function that is 2π periodic in x unless otherwise stated.

3.3.2 A chemical synapse model

An action potential generated in the cell body passes down an axon that terminates in a synaptic bouton. As mentioned earlier, neurotransmitter is then released to interact with the postsynaptic membrane. The neurotransmitter

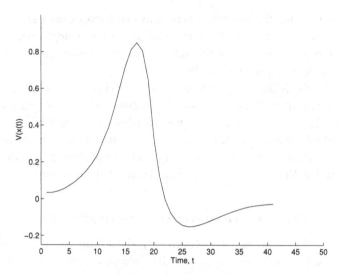

Figure 3.10. A VCON voltage spike.

kinetics at a chemical synapse are analogous to a low-pass filter as shown in Section 3.1.4. The output voltage X is determined by solving the equation

$$RC\frac{dX}{dt} + X = S_X,$$

where in the analogy the time constant for the gap neurotransmitter concentration c, is $RC = 1/(k_{\text{diff}} + k_{\text{post}})$ and S_X describes the source of neurotransmitter.

We suppose that the source of neurotransmitter is proportional to the rate at which the action potentials arrive at the synapse. This is described by the activity of the device, or in terms of the input action potential phase ϕ it is

$$S_X \approx d\phi/dt.$$

The output (postsynaptic impact) of a chemical synapse is taken here to be proportional to the output voltage X, which is determined by solving the equation

$$\chi\dot{X} + X = E_c(\phi),$$

where E_c describes the neurotransmitter release as a function of the input circuit's activity and the time constant χ reflects the kinetic rate constants.

The interaction of neurotransmitters with the postsynaptic membrane is not very well understood [82]. The transmitter can be excitatory, adding to the postsynaptic potential, or it can be inhibitory. We assume that this can be modeled by adding the effect of neurotransmitter to the postsynaptic potential and then trimming the sum to fit the physiological limits of the postsynaptic membrane. This is accomplished in our circuit by combining a voltage adder (+) with a linear amplifier (see Chapter 2). We refer to this circuit as being a *comparison amplifier*.

If X is the output of the synapse, and if U is the postsynaptic potential prior to alteration, then the modified postsynaptic membrane potential is the output of a linear amplifier acting on the sum of U and X if the synapse is excitatory or $-X$ if it is inhibitory. The amplifier output is described by its characteristic function S.

In general, S can be any sigmoidal function (i.e., any monotonically increasing, continuously differentiable function having one inflection point). However, it is often convenient for our simulations to fix notation and take

$$S(u) \approx \tanh u$$

or

$$S(u) = sat(u) = \begin{cases} \operatorname{sgn} u & \text{for } |u| > 1 \\ \sin \frac{\pi u}{2} & \text{for } |u| \le 1. \end{cases}$$

Often in our calculations, we simply take $S(u) \approx u$ when that is convenient.

We note for future reference that if S is a sigmoidal function, then the Fourier series of $S(\cos \theta + \cos \phi)$ has the form

$$S(\cos \theta + \cos \phi) = \sum_{j=1}^{\infty} C_j \cos j(\theta - \phi)$$

$$+ \sum_{j=1, k \ne 0}^{\infty} D_{j,k} \cos \left(j(\theta - \phi) + k(\theta + \phi) \right), \qquad (3.3)$$

where many of the constants $C_j \ne 0$.

3.3.3 An electrical synapse model

We model an electrical synapse in a similar way, but now with the source of neurotransmitter replaced by a term that describes the potential difference between the pre- and postsynaptic potentials.

An action potential ranges from the potassium resting potential of the membrane (roughly -75 mV) to the sodium resting potential (roughly $+55$ mV). An action potential must reach a certain strength before it can drive neurotransmitters due to a threshold effect. Therefore, the first device in the electrical synapse model is a rectifying diode. This takes the positive part of the action potential as being the active part of the signal, so we consider 0 as being the transmitter release threshold. A convenient choice is to take $S(V) = V_+$, where V is the presynaptic potential and $V_+ = \max(V, 0)$ is the positive part of V. The result

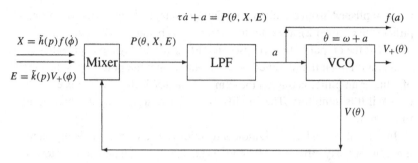

Figure 3.11. VCON circuit.

is that E is determined from the equation

$$\sigma \frac{dE}{dt} + E = E_e(\theta),$$

where E_e is some function of $V(\theta)$. Often we take $\sigma = 0$ and $E = V_+$.

Note that the filters describing these two kinds of synapses have the effect of introducing time delays. This is easy to see from the following calculation: If

$$\tau \dot{U} + U = \exp(i\omega t)$$

then

$$U(t) = U(0) \exp(-t/\tau) + \frac{\exp(i\omega t) - \exp(-t/\tau)}{1 + i\omega\tau}.$$

We see that the decaying terms cannot be considered negligible until $t \approx \tau$. We think of the filter as introducing a time delay of size τ.

3.3.4 The VCON model

Next, we combine these synapse models with our model of the hillock region. We denote by $\dot{\theta} = a + \omega$ the activity of the cell (that is, its firing rate). Thus, a will be described as being the cell's activity, although it is actually a translation by ω different from that. Then, as depicted in Figure 3.11, we have

$$\dot{\theta} = a + \omega, \tag{3.4}$$

$$\tau \dot{a} + a = P(\theta, X, E), \tag{3.5}$$

$$\chi \dot{X} + X = E_c(\phi), \tag{3.6}$$

$$\sigma \dot{E} + E = E_e(\phi). \tag{3.7}$$

Here ϕ is the phase of an input signal arriving through either a chemical or electrical synapse. Multiple inputs will simply add. The function P is critical to the model. It describes the output of a mixer – essentially a phase detector – that combines inputs (X, E) with the cell's state $(V(\theta))$ in an appropriate way.

3.3.4.1 Notation for VCON networks

In describing networks of VCONs, we simply replace the variables a, θ, etc. by appropriate vectors and matrices:

$$\dot{\theta} = a + \omega, \tag{3.8}$$

$$\tau \dot{a} + a = P(\theta, X, E), \tag{3.9}$$

$$\chi \dot{X} + X = E_c(\dot{\theta}), \tag{3.10}$$

$$\sigma \dot{E} + E = E_e(\theta), \tag{3.11}$$

where

$\theta \in T^m, a \in E^m$;
τ, χ, and σ are diagonal matrices of filter time constants;
ω is a vector of center frequencies; and
P is a smooth function, which is periodic in the θ variables.

Systems of this form will be studied in the remainder of this book.

3.3.4.2 First-order scalar VCON and electrical synapses

One useful simplification results in the first-order VCON with electrical synapses. In this case, since $\tau = 0$, the model becomes

$$\dot{\theta} = \omega + P(\cos\theta, E),$$

$$\sigma \dot{E} + E = E_e(\phi),$$

which, when $\sigma = 0$ gives

$$\dot{\theta} = \omega + P\big(\cos\theta, E_e(\phi)\big)$$

Networks of such models represent flows on high-dimensional tori, which are investigated in the next chapter using the rotation vector method.

We derive some results for the first-order electrical model next. The second-order model was analysed in [61].

3.3.5 A free VCON

In the absence of external forcing and with $\tau = 0$, the VCON model becomes

$$\dot{\theta} = \omega + S\big(V(\theta)\big),$$

where S is a sigmoidal function of V. The VCON is said to be excitable if ω

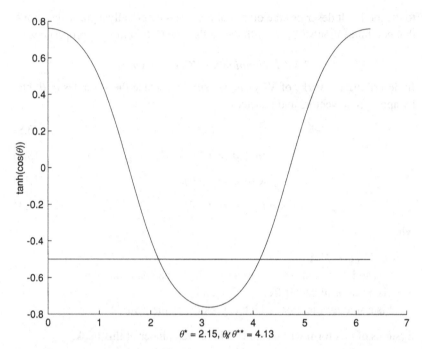

Figure 3.12. Equilibria for VCON when ω is in the range of $S(V(\theta))$. Each crossing of the two curves is a rest point for the phase. Those at which $S(V(\theta))$ is negative are stable, and the others are either unstable or neutral.

is in the range of $S(V(\theta))$. In that case, $\theta \to$ *constant* as $t \to \infty$. Figure 3.12 indicates the situation in this case

For example, if $S(u) = \tanh u$ and $V(\theta) = A \cos \theta$, then we must solve

$$\cos \theta = -\frac{\tanh^{-1} \omega}{A}$$

for θ. This can only be done if

$$\left| \frac{\tanh^{-1} \omega}{A} \right| < 1.$$

In this case, there are two candidates in each interval of length 2π for the limiting value. The arrows in the figure indicate the sign of $\dot{\theta}$, so the lower choice (say θ^*) is the stable one and $\theta \to \theta^*$ (MOD 2π) as $t \to \infty$. The analysis for triangular wave or square wave output V is similar. More complicated choices for V can possibly have several rest points in each interval of length 2π. As ω moves out of the range of $S[V(\theta)]$, a saddle-node bifurcation occurs as discussed earlier.

3.4 Summary

Neuron physiology is complicated. There remain many important and unanswered questions about various mechanisms of action potential generation and propagation and synapse dynamics. The general aspects of neurons used here to construct models and analogue circuits, however, are reasonably well understood.

The Hodgkin–Huxley nerve cell theory continues to provide a significant motivation for the study of neurons. The single-channel-with-escapement models have been quite useful in understanding the full model, and work is continuing on determining their properties.

The VCON model provides some tantalizing possibilities for uncovering phase information about networks, and the remainder of this book deals with some of them. There have been other models based on large networks of transistors, due to Perkel, Lewis, and others, but none has been reduced to the simple, tractable formulation that is exploited here. There has been work by Ermentrout and Kopell, Kuramoto and Storgatz, and others using phase models that appear (partially) as canonical equations for certain bifurcation phenomena, but none is based on constructible electronic circuits like the VCON. Also, we have seen here how various ionic channels can be associated with components of the VCON.

The Hodgkin–Huxley model describes a relaxation oscillator, much like the VCON when V is taken to be a square wave. However, the primary feature of the VCON is that it is given in terms of the output phase. This is important since the analysis of frequency in any model requires the derivation of an equation for the phase, and this is difficult to do in the Hodgkin–Huxley and shunting models. It is for this reason that the remainder of this book studies neural networks using VCONs.

3.5 Exercises

1. a. Reproduce the cusp figures presented here in Figures 3.8 and 3.9.
   b. Draw the trajectory of

   $$\dot{v} = -(v^3 - \cos\theta v + \sin\theta), \quad \dot{\theta} = 1$$

   on the same plot as in part a.
2. *First-order VCON.* A VCON saddle-node on a limit cycle bifurcation. Obtain the bifurcation diagram in Figure 3.13: Show that if $|\omega| < 1$ then there are two static states in each interval of length 2π and otherwise there are none, and the solution increases without bound.

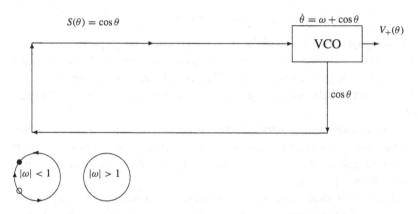

Figure 3.13. Saddle node on a limit cycle: first-order VCON.

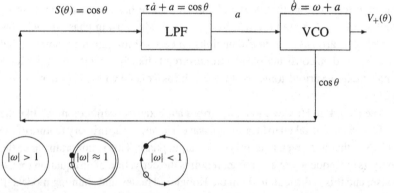

Figure 3.14. Second-order VCON showing coexistence of a saddle (●), node (○), and stable oscillation.

3. *Second-order VCON.* Show the coexistence of a saddle, a node, and a stable oscillation. As in the previous exercise, obtain the bifurcation diagram in Figure 3.14.

4. *A gating model.* Consider excitation of a self-stimulated neuron like that described in this chapter, where the excitatory signal is $I = F(V)$. Here $F(V)$ is a sigmoidal function

$$F(V) = \beta V/(K + V).$$

The membrane model becomes

$$C\dot{V} = \beta z F(V)(E_1 - V) - g_3 V,$$

$$\dot{z} = S - \alpha z F(V).$$

Describe solution behavior of this system under various conditions on the data α, β, S, K, E_1, and g_3. For example, is the solution oscillatory or does it tend to a static state? If the latter, to which static state does it tend?

5. *Analysis of the FitzHugh–Nagumo model.* Consider the FitzHugh–Nagumo model with $R = 0$. Determine where the isoclines cross. Show that, in the three cases described in this Chapter, the static state is stable or unstable.

6. *Firing map for integrate-and-fire model.* Consider the integrate-and-fire model of Section 3.2.6 with $RC = 1$. Let t_n denote a firing time (i.e., $v(t_n) = 1$). If t_{n+1} is the next firing time, derive a formula relating t_n and t_{n+1}.

   Let $S(t) = A + B \cos \omega t$. Write a computer program for finding the resulting firing sequence $\{t_n\}$. Simulate the model by solving it numerically for various values of A, B, and ω. Plot the pairs of points (t_n, t_{n+1}) (MOD 2π) for several different initial points and identify the circle mapping from it.

7. *Free VCON.* Describe the solutions of the free VCON

$$\dot{x} = \omega + \tanh(\cos x)$$

   by plotting $x(t)$ versus t. Assume that $|\omega| < 3/5$ in this calculation.

8. *Bursting of a VCON.* Show that a pulse applied to an excitable VCON can cause a burst of action potentials to be generated. Solve the free VCON model numerically in this case and identify parameters and input voltage amplitudes that will lead to bursting. List how many action potentials are generated by input pulses of various strengths.

4

Signal processing in phase-locked systems

The preceding chapters have introduced several models of neurons and of circuits and have shown how aspects of them can be identified with simple clocks. The connection with simple clocks establishes a bridge for us from physical variables to the frequency domain. Our next step is to determine how such systems in the frequency domain can be studied.

There are well-developed methodologies of signal processing that underlie most of modern telecommunications and computer design. Signal processing includes mathematical methods for determining the frequency and phase responses of devices and the coherence of signals and for comparing signals. Also, it enables us to study the impact of random noise through determination of signal to noise ratios.

The work presented in this chapter relies heavily on Fourier analysis, which we introduce first. Next, we consider the frequency and phase response of various circuits including VCON. The *rotation vector method* enables us to analyze quite general networks, in particular reducible networks and weakly connected networks, for phase locking. Incidentally, the term *phase locking* is used in several ways in the literature. We will see that networks can lock on frequencies, called by some *frequency locking*, and that the phase deviations can lock onto stable values, called here *phase deviation locking*. We include both of these phenomena in our use of phase locking.

The rotation vector method is introduced with the intention of clarifying what phase locking is, what causes it, and how it can be uncovered. There are neural networks that perform a kind of Fourier analysis. One of them is the neural prism that fans an input out onto an array of neurons that are tuned at various frequencies. Their response yields a spatial distribution of signals that presents a decomposition of the input signal onto various Fourier modes. We study such a network next.

The chapter ends with a description of how random noise can enter signals and how it is processed by our circuits. Noise can occur through corruption of signals from a variety of sources, inaccuracies in circuit elements, and imprecise setting of data. In order for noise to be accessible to mathematical analysis, we must make some assumptions about its structure. For this we rely on the Law of Large Numbers, which states (roughly) that processes have a meaningful average value, and the Central Limit Theorem that states that most of the random processes we consider here are closely connected to a variable having a normal (Gaussian) distribution.

4.1 Introduction to Fourier analysis

We refer to "signals." These are simply functions of time, and in applications they might be generated by a mechanical device generating acoustic waves or by electrical devices that generate voltage pulses. However, signals can also occur in numerous other ways that are transduced to our brain through our senses. The signal might carry information, or it might be random noise. If a message can be encoded in a signal and eventually extracted from it, then something useful can be accomplished. For example, in a mechanical control system a signal might cause a mechanical actuator, like the rudder on a ship, to move. Or, in a body, a signal carried by a nerve fiber might cause a particular muscle to contract.

Let us consider a signal, say $f(t)$, that is smooth (say, twice continuously differentiable) and periodic, say $f(t + T) \equiv f(t)$ for all $t \in (-\infty, \infty)$. A remarkable fact derived by Fourier in the first part of the nineteenth century is that such a signal, no matter how complicated, can be expressed in terms of simple trigonometric functions. He showed that constants $a_0, a_1, \ldots, b_1, b_2, \ldots$ can be found so that

$$f(t) = \frac{a_0}{2} + \sum_{n=1}^{\infty} \left(a_n \cos \frac{2\pi nt}{T} + b_n \sin \frac{2\pi nt}{T} \right),$$

where this series converges in some nice sense. A significant part of mathematical research since Fourier has been devoted to studying such series and extensions of these ideas.

We see that each term on the right-hand side of this expression has period T since t increasing by T units takes each of the trigonometric terms through at least one full cycle of 2π. The coefficients in this expansion are given by the formulas

$$a_n = \frac{2}{T} \int_{-T/2}^{T/2} f(t) \cos \left(\frac{2\pi nt}{T} \right) dt \quad \text{for} \quad n = 0, 1, \ldots$$

and

$$b_n = \frac{2}{T} \int_{-T/2}^{T/2} f(t) \sin\left(\frac{2\pi n t}{T}\right) dt \quad \text{for} \quad n = 1, 2, \ldots.$$

The functions $\cos \frac{2\pi n t}{T}$ and $\sin \frac{2\pi n t}{T}$ are referred to as being the *modes of oscillation* of f and the coefficients a_n and b_n are referred to as being the *amplitudes of the nth mode*. Note that the modes increase in frequency (they wiggle faster) as the mode number n increases, and because the series converges, their amplitudes get small.

By using a Fourier series to represent a signal f, we accomplish a transformation of f onto a standard set of modes that we can investigate separately. The transformation

$$f \leftrightarrow \{a_n, b_n\}$$

is referred to as being the *Fourier transform of f*.

Important extensions of this approach can be used when f is not periodic [139] or when f is only known at discrete times [103].

We know from calculus that

$$e^{i\theta} \equiv \cos\theta + i\sin\theta,$$

where $i = \sqrt{-1}$. This observation enables us to write the Fourier expansion of a signal f more economically as

$$f(t) = \sum_{n=-\infty}^{\infty} f_n \exp\left(\frac{2\pi i n t}{T}\right),$$

where

$$f_n = \frac{1}{T} \int_{-T/2}^{T/2} f(t) \exp\left(-\frac{2\pi i n t}{T}\right) dt.$$

This follows from direct substitution in the formulas for a_n and b_n:

$$f_n = \frac{a_n - i b_n}{2} = \frac{1}{T} \int_{-T/2}^{T/2} f(t) \exp\left(-\frac{2\pi i n t}{T}\right) dt, \quad \text{for} \quad n = 1, 2, \ldots.$$

Moreover, if f is a real-valued function, then

$$f_{-n} = \bar{f}_n = \frac{a_n + i b_n}{2}, \quad \text{for} \quad n = 1, 2, \ldots$$

Now, of course, we are dealing with complex numbers for both the amplitudes and the modes of oscillation. The numbers

$$|f_n|^2 = a_n^2 + b_n^2$$

give the amplitudes squared, sometimes referred to as being the power of the nth mode.

These numbers can be computed directly from f in an interesting and useful way. We define the *autocorrelation function* to be

$$\chi(\tau) \equiv \lim_{T \to \infty} \frac{1}{T} \int_0^T f(t + \tau) \bar{f}(t) \, dt.$$

In terms of the Fourier series we see that

$$\chi(\tau) = \sum_{n=-\infty}^{\infty} |f_n|^2 \exp\left(\frac{2\pi i n \tau}{T}\right).$$

Thus the coefficients of the Fourier expansion of the autocorrelation function describe the power of the various modes. That is, the Fourier expansion of χ is simply the one for f with the amplitude replaced by the *power* of each mode. By plotting the values of $|f_n|^2$ against n, we get a quick visual reading of what frequencies are dominant in a signal. Such a graph plots the *power spectrum of f*.

For example, consider the function $f(t) = \cos 2\pi t (1 + 100\cos_+ t)$. The graph and the power spectrum of this function, which resembles one obtained in neuron bursting, are displayed in Figure 4.1.

The cross-correlation function of two signals, say f as above and

$$g(t) = \sum_{n=-\infty}^{\infty} g_n \exp\left(\frac{2\pi i n t}{T}\right),$$

a function also having period T in t, is defined to be

$$\chi_{(fg)}(\tau) = \sum_{n=-\infty}^{\infty} f_n \bar{g}_n \exp\left(\frac{2\pi i n \tau}{T}\right).$$

The cross-correlation function enables us to compare modes in f and g, and there are special neurons in the brain that perform such an operation on two signals. This is zero if f and g have no modes in common and so are uncorrelated.

It is possible to compute all of these processes on signals using packages based on the fast Fourier transform (FFT). These are available in many computer packages.

We will also consider functions that depend on several phase variables. Consider

$$F(\theta_1, \ldots, \theta_N),$$

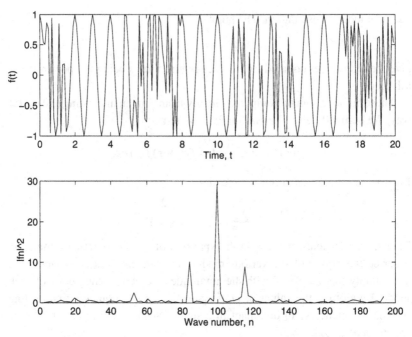

Figure 4.1. Graph of $f(t) = \cos 2\pi t (1 + 100 \cos_+ t)$ vs. t (top), and its power spectrum $|f_n|^2$ vs. n (bottom).

where F is twice continuously differentiable in its variables and is 2π periodic in each of them. Such a function can be expanded in a multivariate Fourier series:

$$F(\theta_1, \ldots, \theta_N) = \sum_{n_1, \ldots, n_N = -\infty}^{\infty} F_{n_1, \ldots, n_N} \exp\left[i(n_1\theta_1 + \cdots + n_N\theta_N)\right].$$

We write this more concisely using standard notation as

$$F(\boldsymbol{\theta}) = \sum_{|\mathbf{n}| = -\infty}^{\infty} F_{\mathbf{n}} e^{i\mathbf{n}\cdot\boldsymbol{\theta}},$$

where $|\mathbf{n}| = n_1 + \cdots + n_N$, and we have used the vector notation

$$\mathbf{n} = (n_1, \ldots, n_N),$$

etc. to emphasize that these are multivariate expansions. In what follows, we will not necessarily use this vector notation in a consistent way for multivariate functions.

The *method of averaging* is based on the following calculation:

$$\frac{1}{T}\int_0^T F(\omega t)\,dt = \frac{1}{T}\int_0^T \sum_{|n|} F_n e^{i(n \cdot \omega)t}\,dt$$

$$= \sum_{n \cdot \omega = 0} F_n + \frac{1}{T}\int_0^T \sum_{n \cdot \omega \neq 0} F_n e^{in \cdot \omega}$$

$$= \bar{F} + \frac{1}{T}\sum_{n \cdot \omega \neq 0} F_n \frac{e^{in \cdot \omega T} - 1}{i(\omega \cdot n)},$$

where $\bar{F} = \sum_{n \cdot \omega = 0} F_n$. It is not clear that the last series in this expansion converges since it can happen that the denominators $(\omega \cdot n)$ can become arbitrarily small. These are called *small divisors*, and they can complicate the analysis of certain problems.

We avoid this problem here since we usually take our approximations to wave forms to be finite Fourier series; these are called *trigonometric polynomials*. This is not made explicit in most cases, but it does enable a rigorous mathematical analysis of our models.

4.2 Frequency and phase response of a VCON

The following example illustrates what we are after here. Consider a phase-locked loop circuit with input phase y described by

$$\dot{x} = \omega + P(x - y),$$

$$\dot{y} = \mu,$$

where the phase detector output $P(u)$ is a periodic function of u, having period 2π. Thus, P can be expanded in a Fourier series

$$P(x - y) = \sum_{n=-\infty}^{\infty} P_n e^{in(x-y)},$$

where the coefficients P_n in this expansion are complex numbers that are determined by the formulas

$$P_n = \frac{1}{\pi}\int_{-\pi}^{\pi} P(u)\,e^{-inu}\,du,$$

where we have substituted $u = x - y$ in the formulas.

We first take the difference of the two equations. This gives

$$\dot{u} = \omega - \mu + P(u). \tag{4.1}$$

We see that if

$$\mu - \omega \in \text{range of } P(u) \tag{4.2}$$

then u will approach a constant, say $u \to u^*$. This has some interesting implications.

First of all, we find that if condition (4.2) is satisfied then

$$\frac{x(t)}{t} = \frac{\mu t + y(0) + u}{t} \to \mu,$$

so we see that the frequency of the response is the same as the driving frequency μ. Since this holds for all data for which (4.2) is satisfied, we say that the system is in *phase lock* over this set of data.

Second, the phase deviation of the response is determined from the following calculation:

$$x(t) - \mu t \to \phi + u^*,$$

where u^* is the limit of u as described above; therefore, the response is determined from the equation

$$0 = \omega - \mu + P(u^*).$$

There may be several choices for solutions of this equation. Each is selected on the basis of where the initial condition $x(0)$ is specified.

A particularly interesting point of view comes from considering the antiderivative of P. Suppose that $Q(u)$ is a function for which

$$dQ(u)/du = \omega - \mu + P(u).$$

The function Q can be obtained from straightforward integration,

$$Q(u) = (\omega - \mu + P_0)u + \sum_{n=-\infty, n\neq 0}^{\infty} \frac{P_n}{in} e^{inu}.$$

The minima of Q correspond to stable choices for u^*, and so we can view the graph of Q as being a memory surface, which we call later a *mnemonic surface*, whose potential wells (minima) correspond to remember states of the system in the sense that if $x(0)$ is in such a well, the phase response of the system will be $\phi + u^*[x(0)]$.

For example, if $P(u) = \sin 2u$, then $Q(u) = (\omega - \mu)u - (\cos 2u)/2$. This is a wiggly ascending (if $\omega > \mu$) line whose minima correspond to stable phase deviations for the response. There will be two minima in each interval of length 2π if condition (4.2) is satisfied.

Only the particular minimum selected by the initial condition for $x(0)$ will appear in the response. Therefore, initializing the system will correctly select one of two distinguishable responses. This shows how information can be encoded in phase deviations by a phase-locked loop; we will discuss this in greater detail and generality later.

The situation for a VCON is quite similar. Consider the model

$$\dot{\theta} = a,$$

$$\tau\dot{a} + a = \omega + \cos\theta + \mathcal{P}(\theta, y),$$

$$\dot{y} = \mu,$$

where the input to the circuit is described by the input phase y and the function $\mathcal{P}$. This input is passed through a low-pass filter having time constant τ. The only part of this signal that will pass the filter is the part that is low frequency. In particular, if we write

$$\mathcal{P}(\theta, y) = P(\theta - y) + R(\theta, y),$$

where R contains higher frequency terms, and if, as before, $\theta - y$ is converging to a constant, then only $\omega + P$ will pass the filter. Therefore, for these purposes, we approximate the system by

$$\dot{\theta} = a,$$

$$\tau\dot{a} + a = \omega + P(\theta - y),$$

$$\dot{y} = \mu.$$

Setting $u = \theta - y$ in this system gives

$$\dot{u} = a - \mu,$$

$$\tau\dot{a} + a = \omega + P(u)$$

or, equivalently,

$$\tau\ddot{u} + \dot{u} - P(u) = \omega - \mu.$$

If $|\omega - \mu|$ is small, then this system has only stable rest points and saddles. In any case, if this condition is met, $u \to u^*$ for some constant u^*. As before, we see that the frequency response of the system is

$$\lim_{t\to\infty} \frac{\theta(t)}{t} = \mu$$

and the phase deviation is

$$\theta(t) - \mu t \to y(0) + u^*.$$

In this manner, we can calculate both the frequency and the phase response of the VCON.

4.3 The rotation vector method

The aim of the rotation vector method is to discover when the vector of phase variables has the special form

$$\theta \rightarrow v\mathbf{1} + u_1\mathbf{W}_1 + \cdots + u_{N-1}\mathbf{W}_{N-1},$$

where the following definitions are used:

- θ is a vector having N components.
- v is a single time-like variable (i.e., $v/t \rightarrow \delta \in (0, \infty)$).
- $\mathbf{1} = (1, \ldots, 1)^{tr}/\sqrt{N}$ is a vector having N components.
- Each $\mathbf{W}_j$ is a vector having N components, and $\{\mathbf{W}_1, \ldots, \mathbf{W}_{N-1}\} \in E^N$ is a collection of vectors that are pairwise orthogonal and that are each orthogonal to $\mathbf{1}$. We suppose that these vectors are normalized so that the matrix $\mathbf{W}$ in $E^{N \times (N-1)}$ whose columns are $\mathbf{W}_1, \ldots, \mathbf{W}_{N-1}$ satisfies:

  a. $\mathbf{W}^{tr}\mathbf{W} = I_{N-1}$, which is the identity matrix in $E^{(N-1) \times (N-1)}$.
  b. $\mathbf{W}^{tr}\mathbf{1} = 0$, $\mathbf{W}_i \cdot \mathbf{W}_j = \delta_{i,j}$.
  c. The matrix $\widetilde{\mathbf{W}}$, which is $\mathbf{W}$ augmented by the column vector $\mathbf{1}$, has determinant 1: $\det(\widetilde{\mathbf{W}}) = 1$.
  d. The components of $\mathbf{W}$ approach 0 as $N \rightarrow \infty$.

- $u_1, \ldots, u_{N-1}$ are scalars.

There are various possible choices for the matrix $\widetilde{\mathbf{W}}$, for example,

$$\widetilde{\mathbf{W}}_{j,k} = \sqrt{\frac{2}{N}} \sin\left(\frac{\pi}{4} + \frac{2\pi jk}{N}\right)$$

is useful. The vectors defined by these components are orthogonal and normalized to have length one. In particular,

$$\det(\widetilde{\mathbf{W}}) = 1,$$
$$\widetilde{\mathbf{W}}^{tr}\widetilde{\mathbf{W}} = \text{identity},$$
$$\widetilde{\mathbf{W}}\mathbf{e}_N = \mathbf{1}, \text{ where } \mathbf{e}_N = (0, \ldots, 0, 1)^{tr} \in E^N, \text{ and}$$
$$\widetilde{\mathbf{W}}^{tr}\mathbf{1} = \mathbf{e}_N.$$

Therefore, the columns of $\mathbf{W}$ form an orthonormal basis for the complement of $\mathbf{1}$ in E^N.

We present in this section two important cases where the rotation vector method can be proved rigorously. These results underlie our computer simulations of more general systems.

4.3.1 Separable case

An interesting case that will play a role in our study of coherence later in this chapter is described by the system

$$\dot{\theta} = \mathbf{1} + \varepsilon \mathbf{\Omega} + P(\theta),$$

where $\theta, \mathbf{\Omega} \in R^N$, P has a special form

$$P(\theta) = \mathbf{W} f(\mathbf{W}^{tr} \theta),$$

and $\mathbf{W}$ is described in the preceding section.

The change of variables

$$\theta = v\mathbf{1} + \mathbf{W}u,$$

where $u \in E^{N-1}$, takes this system to two equations

$$\dot{v} = 1 + \varepsilon \mathbf{\Omega} \cdot \mathbf{1}$$

and

$$\dot{u} = \varepsilon \mathbf{W}^{tr} \mathbf{\Omega} + f(u).$$

Thus, the model can be reduced to two separate problems – one for the time-like variable v and the other for the phase deviations u.

If f is a gradient field, so $f(u) = -\nabla F(u)$, then the system for u is a gradient system

$$\dot{u} = -\nabla\big(-\varepsilon \mathbf{W}^{tr} \mathbf{\Omega} \cdot u + F(u)\big).$$

Minima of the potential function

$$-\varepsilon \mathbf{W}^{tr} \mathbf{\Omega} \cdot u + F(u)$$

correspond to stable equilibria for the phase deviations u. If $u(0)$ is in the domain of attraction of such a minimum, then $u \to u^*$ as $t \to \infty$. As a result,

$$\lim_{t \to \infty} \frac{\theta(t)}{v(t)} = \mathbf{1},$$

and we have that the ensemble is frequency locked in the ratio

$$\theta_1 : \cdots : \theta_N \to 1 : \cdots : 1$$

as $t \to \infty$. Since $v(t)/t \to 1 + \varepsilon \mathbf{1} \cdot \mathbf{\Omega} \equiv v_\infty(\varepsilon)$, the individual frequencies of the ensemble drift with ε, but all of the oscillators are locked to the same frequency [13].

The system is stable in that the phase deviations converge to u^*, and we say that the system is in *phase-deviation lock*. The system being in frequency lock and in phase-deviation lock is referred to as being in *phase lock*. In terms of the oscillator phases, we have

$$\theta \rightarrow v_{\infty}(\varepsilon) t\mathbf{1} + \mathbf{W}u^*(\varepsilon).$$

Thus in the reducible case, we have an interesting result: The ensemble is *coherent*, meaning that the oscillators in it are firing at the same frequency, and information carried by the phase deviations can be decoded.

4.3.2 Highly oscillatory VCON networks

Another system that can be analyzed using the methods derived here is

$$\dot{\theta} = \frac{1}{\varepsilon}\mathbf{1} + \alpha + \cos\theta + P(\theta)$$

where P, α, and θ are vectors having N components and ε is a small positive number. Such networks are referred to as being *highly oscillatory*. We suppose that P is a vector of trigonometric polynomials.

We define new variables by

$$\theta = v\mathbf{1} + Qu$$

where $Q^{tr}\mathbf{1} = 0$. Then

$$\dot{v} = \frac{1}{\varepsilon} + \mathbf{1} \cdot \alpha + \mathbf{1} \cdot \cos\theta + \mathbf{1} \cdot P(\theta)$$

and

$$\dot{u} = Q^{tr}\big(\alpha + \cos\theta + P(\theta)\big)$$

We next write

$$\frac{du}{dv} = \frac{\varepsilon Q^{tr}\big(\alpha + \cos(v\mathbf{1} + Qu) + P(v\mathbf{1} + Qu)\big)}{1 + \mathbf{1} \cdot \cos(v\mathbf{1} + Qu) + \varepsilon\mathbf{1} \cdot P(v\mathbf{1} + Qu)}$$

$$= \varepsilon Q^{tr}\frac{\big(\alpha + \cos\theta + P(\theta)\big)}{1 + \varepsilon(\mathbf{1} \cdot \cos + \mathbf{1} \cdot P)}$$

Next, we average this system with respect to v as discussed earlier. The result is

$$\frac{d\bar{u}}{dv} = \varepsilon Q^{tr}\big(\mathbf{1} + \bar{P}(Q\bar{u})\big) \tag{4.3}$$

where

$$\bar{P}(\bar{u}) = \sum_{n\perp\mathbf{m}} P_n e^{in \cdot Q\bar{u}}.$$

If this system has a stable equilibrium u^*, then the preceding method of analysis can be applied, and we find similar results on frequency and phase-deviation locking.

When will Equation (4.3) be stable? If P has sufficiently many modes so that the equation for $\bar{u}$ has stable equilibria, then the method can be carried through.

4.3.3 Networks with filters in them

The models described in Chapter 3 for the electrical and chemical synapses have the form

$$\dot{\theta} = a,$$

$$\tau \dot{a} + a = \omega + P(a, \theta, X, E),$$

$$\chi \dot{X} + X = f(A),$$

$$\sigma \dot{E} + E = V(\mu t + \psi),$$

where A and $\mu t + \psi$ are input activity and input phase, respectively. Then the second equation in the system becomes

$$\tau \dot{a} + a = \omega + P\big[a, \theta, \tilde{k}(p)f(A), \widetilde{K}(p)V(\mu t + \psi)\big],$$

where the terms $\tilde{k}$ and $\widetilde{K}$ are transfer functions of the chemical and electrical synapses. When the filters are acting, only lower frequencies are allowed to pass, but the same calculations apply in this case.

4.4 Frequency and phase response of parallel networks

An interesting network occurs in the auditory pathway [6]. In this, a signal is fanned out onto an array of neurons, each having a slightly different center frequency. The output is a spatial array of neurotransmitters that are related to the signal's Fourier expansion. We can investigate this using the network depicted in Figure 4.2.

$$\dot{\theta}_j = \omega_j + S\big[\cos \theta_j + A_j \cos_+ (\mu t + \psi)\big]$$

for $j \in \overline{1, N}$, where μ and ψ are the input frequency and phase deviation, respectively, and ω_j is the center frequency. The vector of center frequencies is graded so that

$$\omega_j = \omega_1 + (j - 1)\delta$$

for $j \in \overline{1, N}$, where δ is a small number.

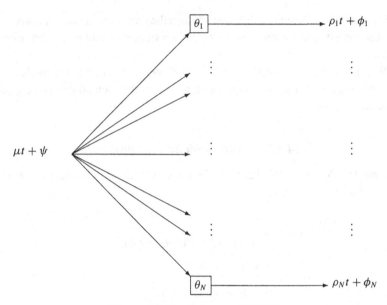

Figure 4.2. Parallel array of VCONs.

As we have seen, each of these isolated VCONs has an output of the form

$$\theta_j \to \rho_j t + \phi_j.$$

The output frequencies resemble the Cantor-function, having steps corresponding to rational numbers and risers over the irrationals. The phase deviations ϕ_j measure appropriate integer combinations of the driving frequency μ and the center frequency ω_j.

The mathematical model for this system, after ignoring what does not get through the filters, is approximately

$$\dot{\theta}_j = \omega_j + A \cos(\theta_j - \mu t - \psi).$$

As a result, we have that

$$\theta_j \to \rho_j t + \phi_j,$$

where these rotation numbers (output frequencies) and phase deviations can be determined by computer simulation. This is illustrated by direct calculation. If $|\omega_j - \mu| \le |A|$ then $\rho_j = \mu$. The phase deviations can be calculated as shown earlier.

Since this network breaks the input function down according to a spectral decomposition, we refer to it as being the *neural prism*. Figure 4.3 shows the frequency and phase response of this network for $\mu = 1.0$ and $\psi = 0.0$.

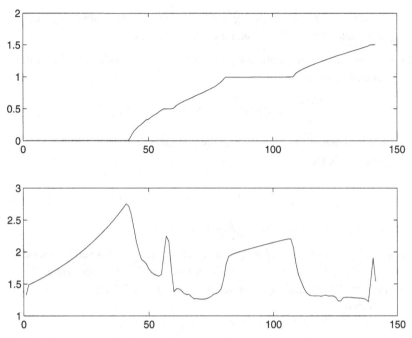

Figure 4.3. In the top plot is shown the output frequency (vertical) vs. the index j. In the bottom plot are shown the phase deviations ϕ_j vs the index j.

4.5 Noise

Noise can arise in networks of neurons through a variety of avenues: The cells have irregularities in their metabolism and in their membrane properties; the chemical bath in which they reside fluctuates; input signals arriving from other cells or from outside the body are irregular; etc. In this section, we present four approaches to studying noise in systems.

4.5.1 Small-amplitude noise

Malkin [97] observed that if a gradient system is perturbed by small-amplitude noise, then its behavior near a minimum is not changed much. This is illustrated by the following example that is relevant to our work on large networks. Consider the system

$$\dot{u} = -\nabla F(u) + \varepsilon g(t, u),$$

where F is a smooth function on E^N and g satisfies the following properties:

g is smooth in u, say its first derivative is continuous;

g is Lebesgue integrable (i.e., $g \in L^1$) in t (so it can be quite wild); and g is bounded uniformly in t and u (so it is not too big!).

Suppose that u^* is a minimum for F. Then for $\varepsilon = 0$, the behavior of u near u^* is described by the equation

$$\frac{dF(u)}{dt} = \nabla F(u) \cdot \left(-\nabla F(u)\right) = -\left|\nabla F(u)\right|^2,$$

which shows that u moves in such a way to minimize F.

When $\varepsilon > 0$, the same calculation shows that

$$\frac{dF(u)}{dt} = \nabla F(u) \cdot \left(-\nabla F(u) + \varepsilon g(t, u)\right)$$

$$= -\left|\nabla F(u)\right|^2 + o(\varepsilon).$$

As a result, u might not tend to u^*, but it will remain within a neighborhood of size $o(1)$ of it (an area that approaches zero as $\varepsilon \to 0$). This calculation shows that in systems with underlying gradient fields, quite wild, but small noise will not disturb the results.

4.5.2 Coherence in the presence of randomly selected data

The study of coupled phase equations having random center frequencies, as might arise in studies of phase-locked loop networks, has been considered by Wiener, Winfree, Kuramoto, and others using ad hoc models. An order parameter has been used to analyze recruitment of oscillators to synchrony, or *coherence* of the network, as the strength of connection increases. Although Kuramoto's work [89] is interesting and useful, it is heuristic, lacks rigorous analysis, and makes some assumptions about symmetry of distributions that might not be necessary for the results [127].

We present here a similar model that accounts for random center frequencies, but one that can be transformed by the rotation vector method into a reducible system that can be analyzed rigorously. With this, we obtain exactly the distribution of coherent oscillators and the distribution of their output frequencies for any level of connection strength.

Consider a network of N phase oscillators in the form

$$\dot{\theta} = \omega^* + F(\theta). \tag{4.4}$$

Suppose:

- ω^* is a vector in R^N of random variables that are identically distributed, each having mean $1/\sqrt{N}$, variance σ^2, and for which $\Omega_j^* = \omega_j^* - 1/\sqrt{N}$ has density function $g(\Omega)$ (that is independent of j). Here and below $*$ indicates a random variable, and **1** is the vector in R^N having all components equal to $1/\sqrt{N}$. This vector is the expected value of ω^*, and it has unit length. Therefore, Ω^* is a vector of identically distributed random variables having identical distributions, whose density is given by the function $g(\Omega)$, and the expected value of Ω, is $E(\Omega_j^*) = 0$ for $j = 1, \ldots, N$.

- Let W_N denote a matrix in $R^{N \times N-1}$ whose columns form an orthonormal basis of the orthogonal complement of **1**. This space is denoted here by $\mathbf{1}^\perp$. The columns of W_N are denoted by $W_{N,j}$ for $j = 2, \ldots, N$, and we assume that

$$W_{N,i} \cdot W_{N,j} \equiv W_{N,i}^{tr} W_{N,j} = \delta_{i,j},$$

where $\delta_{i,j}$ is Kronecker's delta. We denote the components of the matrix W_N by $W_{N,i,j}$ for $i \in \overline{1, N}$ and $j \in \overline{2, N}$. In order to use the Central Limit Theorem later, we suppose that $\lim_{N \to \infty} W_{N,i,j} = 0$.

For example, we can take, as we did earlier,

$$W_{N,k,j} = \frac{1}{\sqrt{N}}\left(\sin\frac{2\pi jk}{N} + \cos\frac{2\pi jk}{N}\right) = \sqrt{\frac{2}{N}}\sin\left(\frac{2\pi jk}{N} + \frac{\pi}{4}\right)$$

for $k \in \overline{1, N}$ and $j \in \overline{2, N}$.

- Let $P_2, \ldots, P_N$ be scalar periodic functions, say having range $[-1, 1]$. For a vector $z \in R^N$ having components $(z_1, \ldots, z_N)$ we denote by $P(z)$ the vector having components $(P(z_1), \ldots, P(z_N))$.

We consider here Equation (4.4) when F has the form

$$F(\theta) = K\sum_{j=2}^{N} W_{N,j} P_j\left(W_{N,j}^{tr}\theta\right) \equiv K W_N P\left(W_N^{tr}\theta\right),$$

where K is a scaling constant.

With these assumptions, the system (4.4) becomes

$$\dot{\theta} = \mathbf{1} + \Omega^* + K W_N P\left(W_N^{tr}\theta\right). \tag{4.5}$$

Kuramoto's system [89], which is based on the choice

$$F_j(\theta) = K\sum_{k=1}^{N} \sin(\theta_k - \theta_j),$$

is convenient for analysis since it is known from theoretical physics how to study order parameters for such systems [26]. The two systems share the following properties:

1. $\mathbf{1}^{tr} F = 0$,
2. $F(\mathbf{1}) = 0$.

The difference between our systems lies in the choice of basis for expressing the coupling dynamics, and in particular our use of the matrix W.

We introduce into Equation (4.5) the change of variables used in the rotation vector method

$$\theta = \mathbf{1}v + Wu.$$

There results a totally disconnected system of N equations:

$$\dot{v} = 1 + \mathbf{1} \cdot \Omega^*, \tag{4.6}$$

in which v appears to be a time-like variable, and for $j = 2, \ldots, N$

$$\dot{u}_j = W_j^{tr}\Omega^* + K P_j(u_j), \tag{4.7}$$

in which the variables u_j appear to be phase deviation variables (lags or leads). We can analyze this system of scalar equations explicitly.

We say that the network (2) is *coherent* (or synchronized or phase locked) if

$$\theta_1 : \cdots : \theta_N \to 1 : \cdots : 1$$

as $t \to \infty$. This will happen if and only if v is timelike and for each $j = 2, \ldots, N$ we have that $u_j = o(t)$. This, in turn, requires that the variables u_j be bounded, which occurs if

$$|W_j^{tr}\Omega^*| \le K.$$

In this case, we say that the phase deviations have been captured or have locked. Other oscillators are said to drift.

The random variables

$$v_j^* \equiv W_j^{tr}\Omega^*$$

have a common density function, say $\tilde{g}(\Omega)$, which can be derived directly from g. Because of the normalization used in defining the vectors $\mathbf{1}$ and $W_{N,j}$, it follows from the Central Limit Theorem [25] that the random variables $\mathbf{1} \cdot \Omega^*$ and v^* approach Gaussian random variables having mean zero and variance σ^2 as $N \to \infty$.

With these preliminaries, we have the following results:

- For any instance of v^*, there are numbers $0 < K_1 < K_2$ such that

    for $K < K_1$ all oscillators drift,
    for $K_1 < K < K_2$ some oscillators drift and some are coherent,
    for $K_2 < K$ all oscillators are coherent.

- The proportion of oscillators that lock is proportional to

$$\int_{-K}^{K} \tilde{g}(v)\, dv,$$

and the proportion that drift with frequency at most L is

$$\left(\int_{-L}^{-K} + \int_{L}^{K} \right) \tilde{g}(v)\, dv.$$

The distribution for this variable (L) resembles the distribution derived in [127 (Figure 5)].

- Since the variables

$$W_{N,j}^{tr}\, \Omega^\star$$

and

$$\mathbf{1} \cdot \Omega^\star$$

approach Gaussian normal variables having mean zero and variance σ^2 as $N \to \infty$, we can deduce the distribution of systems for which v is timelike (i.e., $v \to \infty$ as $t \to \infty$) and for which the phase deviations lock.

There does not seem to be a sharp threshold of K at which coherence almost surely occurs in the limit $N \to \infty$ as reported in [89].

- The variable v is timelike if the instance chosen for $\Omega^\star$ satisfies $\mathbf{1} \cdot \Omega^\star > -1$. That is,

$$v = t(1 + \mathbf{1} \cdot \Omega^\star) \to \infty \text{ as } t \to \infty.$$

- Suppose that $\theta_1, \ldots, \theta_M$ are coherent and the others drift. Then

$$\theta_j/v = 1 + \sum_{k=2}^{N} W_{jk} u_k / v \to 1$$

for $j = 1, \ldots, M$, and

$$\theta_j/v = 1 + \sum_{k=2}^{N} W_{jk} u_k / v \to 1 + \mu_j^\star$$

for $j = M+1, \ldots, N$, where $\{\mu_j^\star\}$ is a set of random variables lying outside the locking interval.

Therefore, we have that

$$\theta_1 : \cdots : \theta_M : \theta_{M+1} : \cdots : \theta_N \to 1 : \cdots : 1 : 1 + \mu_{M+1}^\star : \cdots : 1 + \mu_N^\star$$

as $t \to \infty$.

When t and N are both large, we have that for $i = 1, \ldots, M$

$$\theta_i \approx v + \sum_{j=2}^{N} W_{N,i,j} u_j^*,$$

where

$$K P(u_j^*) \equiv W_j^{tr} \Omega^*.$$

Thus, the frequencies and phase deviations are determined. The coherent oscillators share a common frequency, approximately the average frequency predicted by ω^* (which is nearly one) if N is large, but they have a distribution of phase deviations, and the drifting oscillators have random frequencies and phase deviations.

We have shown here several important aspects of coherence in networks. The model considered here is in the spirit of the one derived and studied in [89]. However, it can be analyzed rigorously since a change of variables transforms it into a totally decoupled system.

As K increases, the proportion of coherent oscillators increases. For sufficiently large K all oscillators are probably (in the technical sense) locked. A distribution of output frequencies similar to Wiener's [138] results from this analysis.

It does not seem to be the case that there is a critical K_c above which all oscillators are probably locked in the limit $N \to \infty$. Rather, the distribution of such K values appears to be normal.

4.5.3 Ergodic random perturbations

A more systematic investigation of the effects of noise on our model is possible. We do not pursue this in depth here other than to indicate the techniques. Details are presented in [62, 122]. We proceed in the following way.

Consider the system

$$\dot{\theta} = \omega\left(y\left(\frac{t}{\varepsilon}\right)\right) + H\left(y\left(\frac{t}{\varepsilon}\right), \theta\right),$$

where:

- $y(t)$ is a Markov random process taking values in a set Y in which it is ergodic with ergodic measure $\rho(dy)$.
- $\theta \in T^N$.
- $\omega(y) \in E^N$ for each $y \in Y$, and it is measurable with respect to ρ.

- ε is a small positive parameter. It measures the ratio of time scales between the noise (fast) and the system's response.
- H is a smooth function of its arguments taking $T^N \to E^N$ for each $y \in Y$, and it is measurable with respect to ρ.

It is shown in [122] that the solution can be expanded using the Law of Large Numbers and the Central Limit Theorem: In particular,

$$\theta = \bar{\theta} + \sqrt{\varepsilon}\theta_1 + \text{ error,}$$

where

$$\dot{\bar{\theta}} = \bar{\omega} + \bar{H}(\bar{\theta})$$

and

$$d\theta_1 = a(\theta_1)\,dt + \sigma(\theta_1)\,dW.$$

We define $\bar{H}$ by the average over Y

$$\bar{H}(\theta) = \int_Y H(y, \theta)\rho(dy),$$

and

$$\bar{\omega} = \int_Y \omega(y)\rho(dy)$$

The variable θ_1 is the limit of the quantity

$$\frac{\theta - \bar{\theta}}{\sqrt{\varepsilon}}$$

as $\varepsilon \to 0$. The nature of convergence of this quantity and the sense in which the error in the formula of the expansion of θ is small are in the sense of distributions over some time interval. Details of this are beyond the scope of this book, and the reader is referred to [122]. However, the behavior of $\bar{\theta}$ is governed by an equation that we have studied earlier in this book. Some insight to the behavior of θ_1 is realized in the next section.

An analysis of this system shows that if the averaged equation is in phase lock, then the stable torus knots defined by their solutions will persist in the presence of such noise in the sense that the rotation vector probably does not change. (See Appendix B.)

4.5.4 Migration between energy wells

Unfortunately, not all noise coming into a system is small or quasi-static as in the preceding discussions. The following calculation shows the influence of rare, but large random perturbations on a system. Consider the stochastic differential equation

$$du = -\nabla F(u)\,dt + \sigma\,dW,$$

where F is a smooth function, $u \in E^N$, and dW describes random noise. W is a Wiener random process, which means that the random variables dW are independent, identically distributed variables whose distribution is standard Gaussian (normal with mean zero and variance 1).

The problem's solution can be described by introducing the probability distribution function that can tell us the probability of the location of a solution at any time. We do this here for the case where $u \in E'$. Rather than expecting to know where the solution of the problem will be at any time as we do in nonrandom problems, we deal with the probability distribution function, which we denote by $\phi(t, u)$ and which we interpret as meaning that the probability that the function $u(t)$ is in a set C is $\int_C \phi(t, du)$. It is known that ϕ solves the equation

$$\frac{\partial \phi}{\partial t} = \frac{\sigma^2}{2}\frac{\partial^2 \phi}{\partial u^2} + \frac{\partial}{\partial u}\big(F(u)\,\phi\big).$$

This problem can be solved for a steady state solution (long-range probability distribution of solutions) by solving

$$\frac{\sigma^2}{2}\frac{\partial \phi}{\partial u} + F(u)\,\phi = 0$$

or, equivalently,

$$\phi(u) = \phi_0 \exp\left(-2 \int_0^u F(u')du'/\sigma^2\right).$$

This equation shows how energy wells for F correspond to modes in the probability distribution, so solutions probably remain near such minima. In fact, it is known [136] that solutions stay in an energy well for a random time proportional to $\exp(-1/\delta)$ where δ = depth of the well. The solution moves among wells at random, but in a way that can be described by a Markov chain. We discussed this from another point of view in Chapter 3.

4.6 Summary

The work in this chapter accomplishes several goals:

Analysis of systems in the frequency domain involves Fourier analysis. The basic ideas of Fourier analysis, in particular expansions, correlations, and spectral properties, are described here. These are basic elements of signal processing, which was studied for many years from the point of view of periodic and almost periodic functions. However, signals that are not periodic can have infinite energy, which complicates analysis. Still, averages of functions of such signals, such as correlations, do make sense and are useful. Once one works with averages of signals, one might as well extend the study to signals that are random functions having certain statistical properties [117], and this provides a firm basis for calculation of a signal's properties, including properties of infinite-energy signals. Although we do not carry this program out in detail here, the ideas of averaging of random signals and of computations using ergodic properties are described.

Methods for recovering the spectrum (center frequencies) of VCON-generated signals and for finding the corresponding phase deviations are also presented in this chapter. In particular, the rotation vector method is introduced. It clarifies the different roles played by the Fourier modes in coupling within a system, the frequencies, and the phase deviations. One result of this is that, for the most part, solutions of a VCON network eventually take the form

$$\theta \to \rho t + \phi,$$

and this provides a basis for some of our work in later chapters.

While knowledge of random processes provides guidance for calculations about signals, it also enables us to study the impact that random fluctuations can have on a system. This is important because there are many variabilities in a real system that are not easily accounted for in models. Therefore, by considering systems in random environments where the randomness has some observable attributes, one can estimate the impact that random fluctuations will have on predictions. Implementation of a theory in applications also involves sampling errors, and usually these become problems for statistics.

4.7 Exercises

1. *The frequency domain.* Let $f(t)$ be a continuously differentiable function that is periodic, say having period T. Show that there is a 2π-periodic function F and a function $\theta(t)$ such that

$$f(t) = F(\theta)$$

and $\dot{\theta} = 2\pi/T$.

2. *Power spectrum.* Find the Fourier series, the correlation function, and the power spectrum of

$$f(t) = (1 + 100\cos t)\cos 2\pi t.$$

3. *Rotation vector analysis.* Carry out the rotation vector analysis for the system

$$\dot{\mathbf{x}} = \omega + \varepsilon\omega_1 + \varepsilon\mathbf{f}(\mathbf{x}, \varepsilon),$$

where

$$\mathbf{f}(\mathbf{x}, \varepsilon) = \sum_{|\mathbf{j}|=-\infty}^{\infty} a_j(\varepsilon)\exp(i\mathbf{j}\cdot\mathbf{x})$$

is the Fourier series of $\mathbf{f}$. Here $\mathbf{x}$, ω, etc. are N-vectors, $\mathbf{j} = (j_1, \ldots, j_N)$, $\mathbf{j}\cdot\mathbf{x} = j_1x_1 + \cdots + j_Nx_N$, and $|\mathbf{j}| = j_1 + \cdots + j_N$. In particular, show that the average of $\mathbf{f}$ with respect to v when $\varepsilon = 0$ is

$$\mathbf{f}^*(\mathbf{U}) = \sum_{|\mathbf{j}|=-\infty}^{\infty} a_j(0)\exp\left(i\mathbf{j}\cdot\sum_{k=2}^{k=N} U_k\mathbf{W}_k\right),$$

where the sum is over all $\mathbf{j}$ for which $\mathbf{j}\cdot\omega = 0$. Show that this system phase locks for any ω_1 that is in the range of $\mathbf{f}^*$.

4. *Phase locking of a multimodal phase-locked loop.* Carry out a phase-locking analysis of the equation

$$\dot{x} = \omega + \varepsilon Ae^{(2ix+3it)} + \bar{A}e^{-(2ix+3it)}.$$

5. Consider the vectors

$$\mathbf{W}_k = \sqrt{\frac{2}{N}}\begin{pmatrix} \sin\left[\frac{2\pi 1k}{N} + \frac{\pi}{4}\right] \\ \sin\left[\frac{2\pi 2k}{N} + \frac{\pi}{4}\right] \\ \vdots \\ \sin\left[\frac{2\pi Nk}{N} + \frac{\pi}{4}\right] \end{pmatrix}$$

for $k \in \overline{1, N-1}$. Show that these vectors are orthogonal to each other and to **1**.

6. Describe the asymptotic behavior of solutions to

$$\dot{x} = 1 + \varepsilon\cos(2x)$$

as a function of their initial position $x(0)$.

7. Given a tolerance δ, show that there are many values of n_1 and n_2 such that $|n_1 + n_2\sqrt{2}| < \delta$ (see [102]).

8. Using a computer, solve the system of differential equations

$$\dot{x} = \omega + \tanh(\cos x + \cos_+ y),$$

$$\dot{y} = v$$

to time $T = 200\pi$ and plot $x(200\pi)/y(200\pi)$ for various values of ω and v.

9. Consider the differential equation

$$\dot{x} = H\big(y(t/\varepsilon), x\big),$$

where

$$H\big(y(t/\varepsilon), x\big) = \cos\big(y(t/\varepsilon) - x\big)$$

and $y(t)$ is an ergodic random process that takes on the value -1 with probability $1/6$, 0 with probability $1/2$, and the value 1 with probability $1/3$. That is, the ergodic measure for y is $\rho(-1) = 1/6$, $\rho(0) = 1/2$, $\rho(1) = 1/3$. Calculate the average

$$\lim_{T \to \infty} \frac{1}{T} \int_0^T H\big(y(t), x\big)\, dt$$

and solve the equation

$$\dot{\bar{x}} = \bar{H}(\bar{x}).$$

Relate your answer to the solution of the original problem using the results described in Section 4.5.3.

5

Small physiological control networks

Neurons, acting together in nuclei, control biological rhythms in our bodies in addition to a variety of other functions including focusing attention and memorization. Most of the discussion of biological rhythms in this book centers around neurons, which we view to be like generators of oscillating currents that drive electrical clocks. Higher-level rhythms will be studied using the methods we derived to study VCONs. In fact, VCONs are quite similar to simple clocks that are used to model higher-level rhythms.

This connection enables us to discuss several examples of higher-level timers and control mechanisms and, through these examples, to illustrate how networks can be constructed, analyzed, and simulated in the frequency domain.

Our bodies include at least two independent timers. These control body temperature and activity/rest cycles. In addition to these, there are faster acting clocks such as heartbeat, respiration, and modulation of blood sugar levels. It is difficult to grasp the complexity of monitoring many clocks running at different rates: It is like watching many clock faces at once and performing some task at specified times determined by the ensemble. Using mathematics, we can visualize multiple clocks as describing a point on a torus whose dimension equals the number of clocks.

Synchronization of body timers is quite common. Biological clocks are rarely isolated since they all operate in a common chemical bath and there are many direct electrical and chemical connections between them. Such interactions between clocks usually lead to synchronization, although synchronizing mechanisms are difficult to uncover. This is even true in physical systems. For example, in 1665 Huygens [71] pointed out that although there was no apparent coupling between several pendulum clocks on the wall of a clock shop, subtle vibrations in the wall and surrounding air tended to synchronize the pendulums.

Our earlier work on VCONs suggests how to model frequency and phase deviations in neural networks and how to determine their responses to external stimulation. In fact, the rotation vector method shows that a collection of clocks that are phase locked behaves like a single clock on one time scale (v) but having multiple outputs. This work carries over to the study of higher-level systems. The examples in this chapter comprise a basic exciter-inhibitor pair that produces regular bursts of activity, a model of shivering and flight in moths, the control of breathing during exercise, rhythm splitting of activity/rest cycles in small mammals, sound localization by binaural animals, and a model of auditory processing.

In the first of these, a model, referred to here as the *atoll model*, produces regular bursts of activity. This is motivated by a network in the thalamus-reticular complex region of the human brain where attention can be focused on one of competing stimuli. This circuit is used in Chapter 7 as part of a network that describes focusing of attention.

The second example describes a network that controls elevator and depressor muscles in moth flight. Part of this is a warm-up period during which the wings shiver.

The third example is von Euler's model for the control of respiration during exercise. This involves a network of three VCONs that describe breathing in, breathing out, and inhibitory feedback from pulmonary stretch receptors. Numerical simulations of this model show breathing patterns similar to those of resting and running humans. In running, the control network responds to periodic loading of the diaphragm, such as various stable synchronizations between breathing and stride that do occur when a person is running. This provides evidence of phase locking within this network. Good runners use several breath/stride ratios during races although 1:2 and 1:1 are the most common.

The fourth example describes how small mammals govern their activity/rest cycles throughout the year. Exercise-wheel activity by a caged vole (e.g., Microtus montanus) shows it to be *nocturnal* when external light/dark (L:D) cycles mimic winter, say L:D = 8:16. (This notation indicates that the vole experiences 8 hours of light and 16 hours of dark each solar day.) When L:D = 16:8, as in summer, the vole is active during the day, i.e., it is *diurnal*. The transition between these two modes of behavior as L:D changes from winter to summer is described in Section 5.5. The changeover from nocturnal to diurnal activity occurs through rhythm splitting. Rhythm splitting by small mammals is apparently governed by a small network of neurons in the suprachiasmatic nuclei, and we show in Section 5.5.4 that two VCONs can reproduce similar behavior.

Next, we introduce a larger network of neurons that describes sound localization structures in binaural animals. In this, an array of neurons is coupled to each of the two ears, and the difference of transmission times measured through

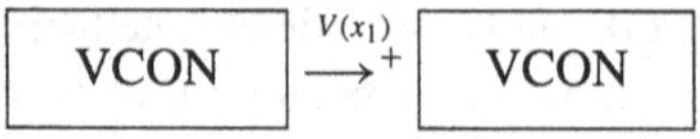

Figure 5.1. Circuit equivalent to $N_1 \to^+ N_2$.

phase deviations between signals in the arrays determines a mapping of where sound originates. Analysis of this network entails calculating transit times of signals in a network.

Finally, a model for signal processing in the auditory pathway, called the tonotope, is presented and studied.

5.1 Notation for neural networks

The notation

$$N_1 \to^+ N_2$$

denotes an excitatory synapse (either electrical or chemical) from one cell to another, that is, from neuron N_1 to neuron N_2. The VCON model of this network with an electrical synapse having no time delay is

$$\dot{x}_1 = \omega_0 + S\big(V(x_1)\big),$$
$$\dot{x}_2 = \omega_0 + S\big(V(x_2) + V_+(x_1)\big),$$

where x_1 is the phase of action potentials at the trigger region of neuron N_1 and x_2 is the action potential phase on neuron N_2. The function S describes how we combine the signals, and for most cases the results are (qualitatively) insensitive to our choice of S, which might be sigmoidal (ogive) or linear as described earlier. The schematic of this network is shown in Figure 5.1.

An inhibitory synapse can be modeled by the same equations but with a sign changed: It is denoted by

$$N_1 \to^- N_2,$$

and the model is

$$\dot{x}_1 = \omega_0 + S\big(V(x_1)\big),$$
$$\dot{x}_2 = \omega_0 + S\big(V(x_2) - V_+(x_1)\big).$$

We will use this notation in describing networks.

The form of V was discussed earlier. Usually we take $V(x) = \cos x$. The superthreshold part of V is denoted by V_+. Conceptually, we take this to be the positive part of $V = \max\{0, \cos x\}$. However, this choice can cause problems for computer simulations since the derivative can be discontinuous at $\cos x = 0$. Therefore, for computer simulations, we take a smoothed function of this $(V_+(x) = \cos x \cos_+ x)$ or we replace V_+ by the first few terms of its Fourier expansion. Recall that $(\cdot)_+$ denotes the positive part of the quantity in the brackets; thus, this represents the rectified voltage.

5.2 The atoll model

In numerous places in the brain two cells interact, one inhibitory and the other excitatory, in such a way that one fires a burst of action potentials followed by the other firing a slow pulse [95, 21, 64, 117]. The model described in this section captures this phenomenon. We study the particular model derived here, called the *atoll model*, and then we use it later in Chapters 6 and 7 in a variety of applications.

The pair of VCONs described by the equations

$$\dot{x} = 5.0\big(1 + \mu + V(x) - V_+(y)\big),$$

$$\dot{y} = 0.05\big(1 + V(y) + 10V_+(x)\big)$$

is referred to as being an *atoll oscillator* since its quasi-static approximation $(\dot{x} = 0)$ resides on an atoll or ringlike set on the torus $\{(x, y) \text{ MOD } 2\pi\}$. μ is a small positive constant, and this system bursts repetitively when $\mu > 0$.

We will study this system in two ways: phase-plane analysis of solutions and computer simulation.

The solutions of this system can be nicely depicted using toroidal coordinates. Consider the square $\{0 \leq x \leq 2\pi\} \times \{-\pi \leq y \leq \pi\}$ as shown in Figure 5.2 (lower right panel).

The solutions move across this square, exiting from one side and entering again at the same point on the opposite side. The solutions slow down and change x direction when they hit the set where $\dot{x} = 0$. This situation is defined by

$$1 + V(x) = V_+(y).$$

Because this is a system of two smooth, doubly periodic equations, Denjoy's [22] theory of rotation numbers applies, and it suggests simulation of the rotation number as described in Chapter 3 by calculating solutions and plotting the rotation number approximation

$$\rho \approx \frac{x(N\pi)}{y(N\pi)}$$

for some large number N. Thus, for some tunings of parameters, we expect regular periodic solutions and for others ergodic solutions.

This system is of the general form

$$\varepsilon \dot{x} = f(x, y),$$

$$\dot{y} = g(x, y),$$

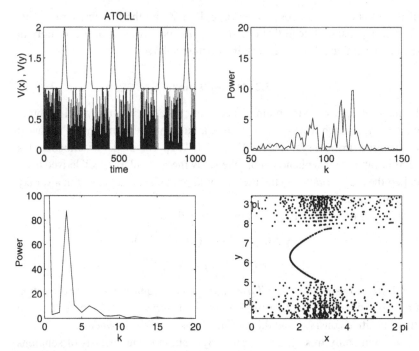

Figure 5.2. Atoll model. This figure shows the phase portrait of a solution (lower right), its temporal evolution (upper right), and its power spectrum ($V(y)$ lower left, $V(x)$ upper right).

where

> f and g are doubly periodic, that is, $f(x+2\pi, y) \equiv f(x, y) \equiv f(x, y+2\pi)$ for all values of (x, y), etc.;
>
> ε is a small positive parameter that describes the ratio of time scales; and
>
> $g(x, y) > 0$ for all (x, y).

This problem can be completely analyzed using methods of ordinary differential equations since it can be reduced to a single equation

$$\frac{dy}{d\xi} = \frac{g(\xi/\varepsilon, y)}{f(\xi/\varepsilon, y)}$$

by taking the ratio of the two equations and setting $\xi = \varepsilon x$. This analysis would benefit from a reasonably straightforward application of the method of averaging were it not for the fact that $f = 0$ at some points encountered by the solutions. Because of these complications, we do not pursue this analysis further here.

It is clear that the rotation number will be of order $1/\varepsilon$. Therefore, in plotting

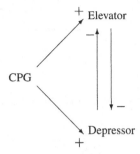

Figure 5.3. Neural network for shiver and flight. A central pattern generator (CPG) drives the elevator and depressor neuron groups. They are reciprocally inhibitory.

our simulations, we would rescale the rotation number by $1/\varepsilon$. The implication of this for the solution's behavior is that whatever our choice for V, $V(x)$ will fire approximately 100 times (a sequence called a burst) for every one oscillation of $V(y)$. The ratio 100 was chosen here since it corresponds to what is observed in the gastric mill controller in lobsters [117]. Simulation of the atoll model with $V(x) = \cos x$ and $\mu = 0.03$ is shown in Figure 5.2.

5.3 Shivering and flight in Hawk moths

It is possible to formulate a minimal VCON model that operates like the neurons known to govern warm-up and flight in insects [78–80]. The following describes the derivation of such a network, its mathematical analysis, and computer simulation. We consider a network of three neurons. The first (labeled 0) is a *central pattern generator* (CPG) that is either off (not firing) or on (firing at a fixed rate). It excites two neurons, one whose firing stimulates contraction of wing elevator muscle (labeled 1) and the other whose firing stimulates the contraction of wing depressor muscle (labeled 2). The network has the form indicated in Figure 5.3.

We let $\theta_0(t)$ denote the phase of the central pattern generator VCON at time t and $\theta_1(t)$ the phase of the elevator and $\theta_2(t)$ the phase of the depressor VCONs at time t. The respective voltage outputs (action potentials) are described by $V_+(\theta_0)$, $V_+(\theta_1)$, and $V_+(\theta_2)$.

These phase variables satisfy the following differential equations:

$$\dot{\theta}_0 = 1.0 + P + V(\theta_0), \tag{5.1}$$

$$\dot{\theta}_1 = 1.0 + V(\theta_1) + V_+(\theta_0) - V_+(\theta_2), \tag{5.2}$$

$$\dot{\theta}_2 = 1.0 + \tau + V(\theta_2) + V_+(\theta_0) - V_+(\theta_1). \tag{5.3}$$

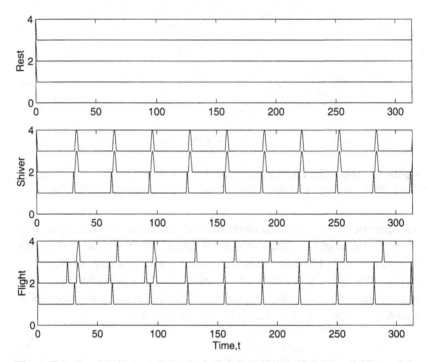

Figure 5.4. Case I (top): rest, $P = \tau = 0$. Case II (middle): shiver, $P = 0.05, \tau = 0.0$. Case III (bottom): flight, $P = 0.05, \tau = 0.08$. In each plot the CPG is shown at the bottom, with the elevator nucleus above it and the depressor nucleus plotted at the top.

The first equation is for the central pattern generator. In it P describes the activation level. When $P = 0$, θ_0 approaches a constant for which $V_+(\theta_0) = 0$, so the CPG is off and there is no output from this device. When $P > 0$, $\theta_0 \to \infty$ as $t \to \infty$, and the device fires repetitively at a constant rate.

The elevator nucleus of neurons is described by the second equation, and the depressor nucleus is described by the third equation. The parameter τ in the third equation describes an activation level of the neuron. When $\tau = 0$, the cell is passive and will fire only when stimulated exogenously to do so. When $\tau > 0$, the cell will fire repetitively.

The three simulations in Figure 5.4 describe cases of rest, warm-up, and flight firing patterns for these VCONs.

The VCON network in Equations (5.1)–(5.3), which involves a driving CPG VCON and two reactive VCONs, can produce firing patterns similar to those observed in Lepidoptera neural dynamics as shown in [80].

In this model we have set two phenomenological parameters (P and τ) to fixed values. The thinking is that these parameters are set by other mechanisms in the body. A decision is made somewhere to turn on the flight CPG, and it is implemented in ways not accounted for in this model (P is set from 0.0 to a positive value). The parameter τ reflects the level of warm-up of the system, and in this minimal model it is set either to 0.0 or to a positive value. It is easy to allow τ to change dynamically as a result of wing neuron activity or muscle temperature. For example, we could take τ to be the filter voltage given by

$$\dot{\tau} + \tau = V_+(\theta_0) + V_+(\theta_1) + V_+(\theta_2)$$

in which τ increases as a result of any activity in the controller neurons – faster if they fire simultaneously. It relaxes to a lower value (cools) in the absence of activity, as would occur if P is set to 0.0 (i.e., when the CPG is shut off [34, 78, 80, 143]).

Shivering is a common mechanism in bodies to raise temperature, and the model, though very simple, lays a basis for modeling shivering in larger systems. See, for example, [29–32] and [128].

5.4 Respiration control

Human respiration during exercise is controlled by neurons that cause muscles attached to the diaphragm to contract. There are muscles for breathing in (inspiration) and for breathing out (expiration). In addition, there are mechanoreceptors (MRs), including pulmonary stretch receptors, that feedback to inhibit the inspiratory neurons controlling the diaphragm muscles. This model of a pattern generator for respiration was introduced by von Euler [24], and it was not intended to account for all aspects of respiration control.

The diaphragm and the central pattern generator are described in this section using VCONs, and then the model is modified to account for periodic loading of the diaphragm as during running. Some numerical simulations of breathing at rest and while running are presented at the end of the section. We will see that a VCON circuit of this kind can produce regular respiration patterns and that it phase locks to oscillatory forcing, as is observed in runners. The VCON circuit constructed here gives a stable controller of a linear mass-spring system.

5.4.1 The diaphragm

The diaphragm is modeled as being a linear mass-spring mechanical system having friction and a restoring force. If z denotes the deflection of the system from rest, then z satisfies the equation

$$m\ddot{z} + r\dot{z} + kz = F,$$

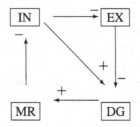

Figure 5.5. von Euler's circuit.

where m is the mass of the system, r is a coefficient of resistance, k measures the restoring force, and F is the external force applied to the system. The external force is applied to the diaphragm by muscles receiving signals from inspiratory and expiratory neurons.

It might be surprising to learn that the stomach and liver, among other things, are attached to the diaphragm. Therefore, the effective mass of the diaphragm changes significantly, especially after eating or drinking. The values of m, r, and k are difficult to estimate, but we choose them for convenience in sample numerical simulations in this section.

An electrical circuit equivalent to the linear mass-spring system is an *RLC* circuit. This is shown in Figure 1.6 where the electromotive force is replaced by an external force.

We take the external force due to muscle contraction in response to inspiratory and expiratory excitation to be

$$F \sim (\text{inspiratory potential})_+ - (\text{expiratory potential})_+.$$

Again, we have used the notation $(\cdot)_+$ to mimic a threshold of an action potential, which when exceeded activates muscles.

5.4.2 A central pattern generator

Figure 5.5 describes a simple analogue of von Euler's respiration control network. The mathematical (VCON) model for this network is

$$\dot{x}_{in} = \omega_0 + \alpha_{in} + AV(x_{in}) - AV_+(x_{MR})$$

$$\dot{x}_{ex} = \omega_0 + \alpha_{ex} + AV(x_{ex}) - AV_+(x_{in})$$

$$\dot{x}_{MR} = \omega_{MR} + AV(x_{MR}) + z_+$$

$$m\ddot{z} + \dot{z} = -kz + BAV_+(x_{in}) - BAV_+(x_{ex}).$$

The notation used in this model is described in the Table 5.1.

Table 5.1. *Data for von Euler's model simulation*

$$\omega_0 = 1.0$$
$$A = 1$$
$$\alpha = 0.5$$
$$\omega_{MR} = 0.01$$
$$S = \tanh$$
$$m = 50$$
$$k = 1$$
$$E = 10\cos t$$
$$B = 50$$

The data ω_0 and α are chosen so that the inspiratory and expiratory neurons are repetitive and the MR neuron is excitable. The synapse characteristic S is taken to be $S(v) = \tanh v$, as in Chapter 1.

The respiration control network in a human body is more complicated than von Euler's model might suggest. First, the diaphragm is a three-dimensional object, and the muscles causing contraction lie on it and in the rib cage. In addition, there are significant nonlinear restoring forces acting on it. There are also a variety of other stretch receptors, chemoreceptors, and baroreceptors that act to modulate breathing. However, this simple model is quite tractable: The numerical simulation in Figure 5.6 produces a realistic breathing pattern and shows that von Euler's mechanism can act as a pattern generator as observed for breathing during exercise.

5.4.3 Respiration while running

Running causes a significant periodic loading on the diaphragm, since various weights (e.g., the stomach) are attached to it. Running can be accounted for in the respiration model by introducing an external forcing of the diaphragm model: Let $p(t)$ denote the external force due to running. Then the equation for z becomes

$$m\ddot{z} + r\dot{z} + kz = F + p(t).$$

It is quite interesting to study phase locking of this network's responses to external forcing.

Bramble and Carrier [12] have collected data on breathing patterns of running mammals. They showed that when the number of breaths per stride are recorded, various phase-locked combinations are seen to occur at various running speeds. They observed that untrained runners have little synchronization

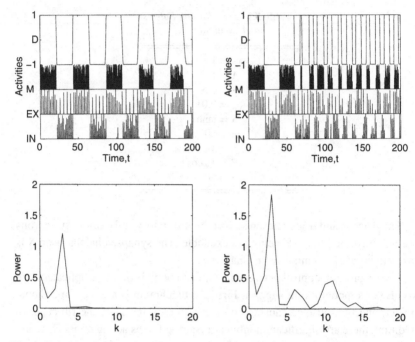

Figure 5.6. Simulations of von Euler's circuit. Left: free respiration (no forcing); Right: forced respiration. Data used are listed in Table 5.1.

between breathing and stride. In contrast, trained runners commonly use the ratios 1:4, 1:3, 2:5, 1:2, 2:3, and 1:1 when running on various grades and at various speeds. Backpackers commonly use 2:1 and 1:1 when hiking.

Breath/stride phase locking can be studied using the respiration control network of the previous section, but with a periodic loading on the diaphragm. To do this, we consider the full system with the z equation replaced by

$$m\ddot{z} + r\dot{z} = -kz + BV_+(x_{\mathrm{in}}) - BV_+(x_{\mathrm{ex}}) + p(t),$$

where p is a periodic function that describes the loading of the diaphragm due to the up and down motion of the internal organs while running.

Experimental observations are in terms of breaths/stride, so we must determine this quantity from the mathematical model to compare it to observations. One method involves introducing polar coordinates in the equation for z: First, we write

$$\dot{z} = (u - rz)/m,$$

$$\dot{u} = -kz + F.$$

Next, we rescale the problem by setting

$$z = Z, \quad u = U\sqrt{km}.$$

The result is

$$\dot{Z} = \omega U - rZ/m,$$

$$\dot{U} = -\omega Z + F/\sqrt{km},$$

where $\omega = \sqrt{k/m}$. Finally we introduce polar coordinates by the change of variables

$$Z = \rho \cos\phi, \quad U = \rho \sin\phi.$$

This results in the equations

$$\dot{\rho} = -(r/m)\rho \cos^2\phi + \left(F/\sqrt{km}\right)\sin\phi,$$

$$\dot{\phi} = -\omega + \left(F/\rho\sqrt{km}\right)\cos\phi - (r/m)\sin\phi\cos\phi.$$

This system of equations is equivalent to the original one for z.

The variable ϕ is the phase of the diaphragm, and if ω_R is the frequency of forcing due to running, then the ratio of breaths to stride is proportional to $\phi(T)/(\omega_R T)$ for large T. This computation is not carried out here, but it is described in the exercises.

As an alternative, we can simulate the model, then calculate the power spectrum of the diaphragm and compare its peaks to ω_R. This is done in Figure 5.6 for $\omega_R = 1$.

5.4.4 Numerical simulation of the respiration model

Figure 5.6 shows the numerical solution of this respiration system with data used earlier. The data are chosen so that in the absence of excitation or inhibition, the in and out neurons fire repetitively at the same frequency, but the mechanoreceptor is excitable.

The right-hand plots are for running, and those on the left are for free (unforced) respiration. The upper plots describe the dynamics in the neurons and the diaphragm; the lower ones describe the power spectrum of the diaphragm in each case.

The bottom signal of each upper plot shows the behavior of the inspiratory neurons (x_{in} MOD 2π), above that is the expiratory neuron's phase, and the plot marked M shows the mechanoreceptor phase. The top plot in the figures shows $\tanh z$, which conveniently normalizes the diaphragm's deviations to fit one scale. When the diaphragm is plotted full scale, one sees that the respiration is quite irregular in amplitude, with alternating periods of deep and shallow breathing. Regular breathing does reappear. The shallow breathing is quite similar to apnea observed in resting adults and more dramatically in infants (see Exercise 1).

In Figure 5.6 (right), a stride is defined to be the time from left footfall to left footfall, and a breath is from inhalation to the next inhalation. The data in this case give a 1:1 ratio of breath to stride.

The diaphragm can be modeled as a (hard) nonlinear spring [58] – the same simulation as described in Figure 5.6. However, such a nonlinear spring shows regular phase-locked respiration during running and there is stronger modulation of the diaphragm's deflection at each stride. The nonlinear spring model enriches the system's phase locking (see Exercise 2).

5.5 Rhythm splitting behavior

This section is divided into two parts. The first summarizes some results from the biological literature. The second presents a VCON network that mimics rhythm splitting.

5.5.1 Some biological experiments in circadian rhythms

As mentioned in earlier chapters, phase-resetting experiments make important connections between the theory and facts of biological rhythms. A set of experiments is described next.

5.5.1.1 Free-running behavior

A typical rhythm experiment is to place an animal, say a hamster or a vole, in a cage equipped with an exercise wheel. This cage is kept in a room where light and temperature can be carefully controlled. A light:dark (L:D) regimen is established in the room by a timing device. Under a 12:12 L:D cycle (i.e., 12 hours of light and 12 of dark each 24-hour day), hamsters are nocturnal; they use their exercise wheel during the dark period. When lights are turned off permanently (D:D), they become free running, with their activity periods starting earlier each day. This pattern is depicted in Figure 5.7. In the following figures, "-" or a blank denotes no activity and "." or "x" denotes activity. Two days are plotted side by side to show clearly what is happening at midnight. The second day is repeated as the first day on the next line. This shows that the animal's free-running period of activity is slightly less than 24 hours.

```
1234567890123456789012301234567890123456789012301 Hour
dddddd1111111111111dddddddddddddd1111111111111dddddd
xxxxx------------xxxxxxxxxxx------------xxxxxx L:D=12:12
xxxxx------------xxxxxxxxxxx------------xxxxxx
xxxx-------------xxxxxxxxxxx------------xxxxxx
xxxxx------------xxxxxxxxxxx------------xxxxxx
xxxxx------------xxxxxxxxxxx------------xxxxxx
xxxxx------------xxxxxxxxxxx------------xxxxxx L:D=0:24
xxxx------------xxxxxxxxxxxx------------xxxxxxx
xxxx------------xxxxxxxxxxxx------------xxxxxxx
xxx------------xxxxxxxxxxxx------------xxxxxxxx
xxx------------xxxxxxxxxxxx------------xxxxxxxx
xx------------xxxxxxxxxxxx------------xxxxxxxxx
xx------------xxxxxxxxxxxx------------xxxxxxxxx
x------------xxxxxxxxxxxx------------xxxxxxxxxx
x------------xxxxxxxxxxxx------------xxxxxxxxxx
------------xxxxxxxxxxxx------------xxxxxxxxxxx
------------xxxxxxxxxxxx------------xxxxxxxxxxx
-----------xxxxxxxxxxxx------------xxxxxxxxxxx-
-----------xxxxxxxxxxxx------------xxxxxxxxxxx-
```

Figure 5.7. Typical free-running activity record for hamsters (mesocricetus spp.). @ denotes the midnight hour.

Such behavior is approximately one activity period per day. We will see cases of *crepuscular activity*, where there are two periods of activity each day, and *ultradian activity* where many short periods of activity are observed throughout the day. Ultradian activity is typical of small animals that have high metabolic rates, such as shrews.

5.5.1.2 Phase-response curves

When hamsters that are in the free-running cycle under constant darkness are exposed to a brief period of bright light, they shift their activity period. In Figure 5.8, we see that when light is applied for a period (denoted by #), the pattern changes: In this case, the animal has delayed its activity. Presumably this is done because it "thinks" that it started during daylight, which is a bad strategy for avoiding predators. The next case (Figure 5.9) shows what happens when the pulse is applied later in what the animal thinks is night, called the subjective night. In this case, we see that the animal started his activity earlier, perhaps thinking that he overran into daylight.

Experiments such as these are summarized in phase-response diagrams. One plot for the golden hamster is shown in Figure 5.10.

Recall the phase-resetting experiments in Chapter 2. The response curve there is quite similar to the one in Figure 5.10. The RIC model predicts a response curve that looks sinusoidal. In fact, a weak pulse applied to a real animal during its subjective day has little or no effect, but the RIC model predicts a delay. The VCON model does correctly account for this quiescent period.

```
1234567890123456789012301234567890123456789012301 Hour
dddddd11111111111ddddddddddddd111111111111ddddddd
xxxxx------------xxxxxxxxxxx------------xxxxxx L:D=12:12
xxxxx------------xxxxxxxxxxx------------xxxxxx
xxxxx------------xxxxxxxxxxx------------xxxxxx
xxxxx------------xxxxxxxxxxx------------xxxxxx
xxxxx------------xxxxxxxxxxx------------xxxxxx
xxxxx------------xxxxxxxxxxx------------xxxxxx
xxxx------------xxxxxxxxxxx------------xxxxxxx
xxxx------------xxxxxxxxxxx------------xxxxxxx
xxx------------xxxxxxxxxxx------------xxxxxxxx
xxxx------------xxx#xxxxxxx------------xxxxxxxx
xxxx------------xxxxxxxxxxx------------xxxxxxxx
xxxxx------------xxxxxxxxxxx------------xxxxxx
xxxx------------xxxxxxxxxxx------------xxxxxxx
xxxx------------xxxxxxxxxxx------------xxxxxxx
xxx------------xxxxxxxxxxx------------xxxxxxxx
xxx------------xxxxxxxxxxx------------xxxxxxxx
xx------------xxxxxxxxxxx------------xxxxxxxxx
xx------------xxxxxxxxxxx------------xxxxxxxxx
```

Figure 5.8. Pulse early in subjective night.

```
1234567890123456789012301234567890123456789012301 Hour
dddddd11111111111ddddddddddddd111111111111ddddddd
xxxxx------------xxxxxxxxxxx------------xxxxxx L:D=12:12
xxxxx------------xxxxxxxxxxx------------xxxxxx
xxxxx------------xxxxxxxxxxx------------xxxxxx
xxxxx------------xxxxxxxxxxx------------xxxxxx
xxxxx------------xxxxxxxxxxx------------xxxxxx
xxxxx------------xxxxxxxxxxx------------xxxxxx
xxxx------------xxxxxxxxxxx------------xxxxxxx
xxxx------------xxxxxxxxx#x------------xxxxxxx
xx------------xxxxxxxxxxx------------xxxxxxxxx
xx------------xxxxxxxxxxx------------xxxxxxxxx
x------------xxxxxxxxxxx------------xxxxxxxxxx
x------------xxxxxxxxxxx------------xxxxxxxxxx
------------xxxxxxxxxxx------------xxxxxxxxxxx
------------xxxxxxxxxxx------------xxxxxxxxxxx
----------xxxxxxxxxxx------------xxxxxxxxxxx-
----------xxxxxxxxxxx------------xxxxxxxxxxx-
----------xxxxxxxxxxx------------xxxxxxxxxxx--
----------xxxxxxxxxxx------------xxxxxxxxxxx--
```

Figure 5.9. Pulse late in subjective night.

5.5.2 *Rhythm physiology*

Many rhythms in mammals are circadian (approximately one day). For example, body temperature, hormone concentrations in blood, and electrolyte excretion in urine are all circadian phenomena. Current speculation is that there are various endogenous rhythm generators in the body, some connected to exogenous signals through light or temperature, some connected to other internal systems

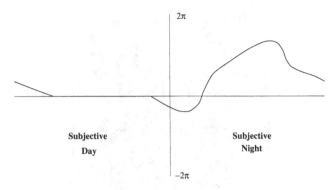

Figure 5.10. Phase response curve for the golden hamster.

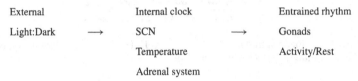

Figure 5.11. Some endogenous, exogenous, and entrained rhythms in humans.

that are entrained by them. The diagram in Figure 5.11 shows some things that are known. Here SCN refers to two collections of neurons, called nuclei, that are located in the brain above the chiasm; hence, these are called suprachiasmatic nuclei. There is evidence that there are at least two endogenous clocks.

5.5.3 Rhythm splitting

Some biologists believe that there are two timers in hamsters, one governing dusk behavior and the other dawn behavior. The data in Figure 5.12 show rhythm splitting of activity patterns in response to a change in photoperiod (i.e., changing from L:D = 18:6 to L:D = 12:12).

5.5.4 A VCON circuit that mimics rhythm splitting

A VCON having two modes, but whose amplitudes change slowly with time, can be constructed that mimics the rhythm splitting described above. Although there is no known connection between the model presented here and the biology of rhythm splitting, this example shows that modulation of a single network element can produce rhythm splitting behavior.

The mathematical model for this circuit is

$$\dot{x} = 1.0 + C \sin \left(\omega^* t - (x/2) \right) + A \sin (\omega^* t - x).$$

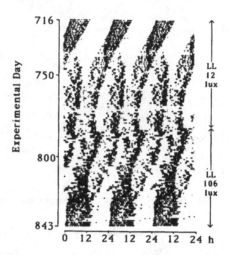

Figure 5.12. Dramatic split data. (Reprinted from K. Hoffman, Biochronometry, Nat. Acad. Sci., Washington, 1971 with permission.)

The phase-locking analysis carried out earlier applies here as well. Three simulations of this model are shown in Figure 5.13. In (a), the model is allowed to run free ($C = 0.2$, $A = 0$). In the second simulation (b), the model is pulsed with $C = 0.2$, $A = 1.1$ for the time interval indicated from 100 to 150. We see that the activity in this simulation has fallen behind the free-running case, and we say that a phase deviation has occurred. In the third simulation (c), the circuit is slowly modulated by setting $C = 0.9$ and

$$A = 1.1 \cos(48t/71).$$

Some biological interpretation can be assigned to these clocks, say $x_{in} = \omega^* t$ corresponds to the phase of gonadal activity that modulates x, the phase of exercise activity. There is no known connection between this circuit and any known organism. We simply give these names to the variables by analogy with small mammals. The clocks can be easily described by stating that when $\cos x_{in}$ is positive, then it is dark; when negative, then it is light. Similarly, when $\cos x$ is positive, there is exercise-wheel activity; when negative, there is none.

In Figure 5.13(c), we see that the activity pattern begins with one activity period in each 24-hour period ($A = 1.1$), changes to two or more active periods ($A \approx 0$), and finally to one period again, but this period is 12 hours out of phase with the original ($A = -1.1$)!

5.6 Sound location by binaural animals

T. N. Parks and E. W. Rubel [105] suggest a mechanism for how chickens can locate sounds by employing a simple neural network. The circuit in Figure 5.14

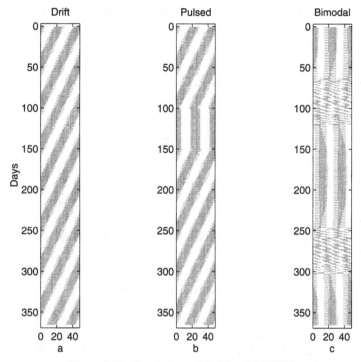

Figure 5.13. Simulations of the bimodal VCON.

$$N \leftarrow N \leftarrow N \leftarrow N \leftarrow N \leftarrow N \leftarrow \text{Left Ear}$$
$$\downarrow \quad \downarrow \quad \downarrow \quad \downarrow \quad \downarrow \quad \downarrow$$
$$N \quad N \quad N \quad N \quad N \quad N$$
$$\uparrow \quad \uparrow \quad \uparrow \quad \uparrow \quad \uparrow \quad \uparrow$$
$$\text{Right Ear} \rightarrow N \rightarrow N \rightarrow N \rightarrow N \rightarrow N \rightarrow N$$

Figure 5.14. Neural network for sound location. N denotes a neuron; the arrows indicate excitatory synapses.

describes essential features of the auditory nuclei of the chicken. In this model, each ear has several axons passing from it, and each arrives at a different central neuron. When one of the central neurons receives simultaneous stimulation from both the right and left sides, it fires. If only one side excites it, there is no firing. The signal from a firing central neuron causes a muscular response, and the chicken responds to where the sound comes from by turning its head.

5.6.1 An axon model

It takes time for an action potential to move down an axon, and the time for the neurotransmitters to diffuse across the synaptic gap has been modeled here by time delays in filters. These two features are important for many reasons. The diffusion process will be ignored here. Axon propagation time, however, turns out to have its own important aspects.

There are a number of ways that axons can be modeled as electrical transmission lines. We introduce a new one here based on the VCON model. In this model of an axon, each VCON acts like a patch of axon membrane, and the coupling along the line is modeled by low-pass filters, say

$$\rightarrow LPF \xrightarrow{z_j} \text{VCON} \xrightarrow{x_j}.$$

All of the VCONs in Figure 5.14 are assumed to be excitable. A super threshold voltage pulse introduced at the left passes through the circuit and eventually reaches the right end. Each VCON introduces a new phase deviation, or time delay, since it takes some time for its output to reach the threshold level of the following VCON. This can be estimated directly from the phase plot of VCON output.

5.6.2 A binaural network

An acoustic wave entering an ear causes the basilar membrane to vibrate in a region that corresponds to the frequency of the acoustic wave. In response, hair cells attached to that region are stimulated to fire. The resulting signal propagates along an axon.

We lump the entire process from the basilar membrane to the nuclei into a single VCON.

An acoustic wave hitting the right ear first results in the right and left ear signals meeting at a receptor cell [104] that is closer to the left ear. Presumably this receptor cell causes muscles to contract, which rotates the head to the right. Exercise 7 outlines an analysis of this network, which we model in the following way using VCONs:

$$\dot{x}_0 = \omega_0 + V(x_0) + L(t),$$

$$\ddot{x}_j + \dot{x}_j = \omega_j + V(x_j) + V(x_{j-1}),$$

$$\ddot{y}_j + \dot{y}_j = \mu_j + V(y_j) + V(x_j) + V(z_j),$$

$$\dot{z}_j = \omega_j + V(z_j) + V(z_{j-1}),$$

$$\dot{z}_0 = \omega_0 + V(z_0) + R(t),$$

for $j = 1, \ldots, N$. A pulse entering the right side followed by one entering the left leads to pulses moving along the respective chains, and the central neuron residing at the point where the two waves are maximum generates the strongest output signal.

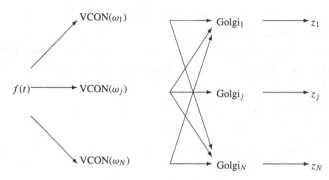

Figure 5.15. The tonotope.

This example demonstrates that temporal patterns of activity in a real neural network do result in observable actions. It clearly demonstrates that brief synchronization of signals is one mechanism used by a brain for processing information.

5.7 The tonotopic mapping in audition

The auditory system in mammals converts an acoustic signal into spatial patterns of neuron firing. These patterns are processed by various auditory nuclei that extract various data, and the emerging patterns of neuron firing are carried to the brain.

Presented here is a VCON network that accounts for two important aspects of auditory processing. Major components of the auditory pathway [6, 55, 57, 104, 106, 131] are:

1. the basilar membrane, whose oscillations are transduced to nerve-fiber firing by hair cells;
2. auditory processing nuclei (e.g., the dorsal cochlea, lateral lemniscus, and the medial geniculate body) that sharpen the auditory signals; and
3. the auditory cortex.

The VCON network studied here accounts for two important aspects of auditory processing nuclei. First, it converts a temporal signal into a spatial pattern of firing, and second, it converts this into a firing pattern that selects only those outputs that are phase locked. The first conversion is accomplished by fanning out the temporal signal onto a linear array of VCONs that are graded by center frequencies as in the neural prism in Chapter 4 (See [63]), so that various VCONs will lock onto particular oscillatory modes of the input signal. The second conversion is accomplished by passing the output of this graded field into an array of VCONs that correlate inputs. We refer to this circuit as being a tonotope [106]. (See Fig. 5.15.)

A signal, say $f(t)$, comes to the tonotope as a series of action potentials just as does a signal coming into processing nuclei from earlier stages in the pathway. $f(t)$ is fanned out onto an array of VCONs that are graded by characteristic firing frequencies, say

$$\omega_0 > \omega_1 > \omega_2 > \cdots > \omega_{N-1} > 0,$$

respectively. The output phases of these VCONs are determined by solving the differential equations

$$\dot{x}_j = \omega_j + S\big(V(x_j) - f_+(t)\big)$$

for $j = 0, \ldots, N - 1$. Emerging from this array are N signals

$$V_+(x_0), V_+(x_1), \ldots, V_+(x_{N-1}),$$

respectively. These pass forward to an array of correlating VCONs.

A correlating VCON is one that uses a product of input signals to control a VCO, but has no feedback, as shown in the graph

$$V_+(x_k)V_+(x_j) \twoheadrightarrow \text{VCO} \to V_+(z).$$

The mathematical model for this correlating VCON is

$$\dot{z} = \mu + \cos_+ x_j \cos_+ x_k.$$

Therefore, we see that the output frequency, or mean activity, of the correlating VCON is

$$\frac{z(t)}{t} = \mu + \frac{z(0)}{t} + \frac{1}{t}\int_0^t \cos_+ x_j(s)\cos_+ x_k(s)\,ds \equiv \omega_z,$$

which as $t \to \infty$ approaches the correlation between the input signals plus the bias μ.

The mathematical model for the tonotope is

$$\dot{x}_j = \begin{cases} \omega_j + S\big(V(x_j) - f_+(t)\big), & \text{for } j = 0, \ldots, N-1, \qquad (5.4) \\[2mm] \mu + V_+(x_{j-N}) \displaystyle\sum_{k=0}^{N-1} A_{k,j-N} V_+(x_k), & \text{for } j = N, \ldots, 2N-1, \end{cases}$$

$$(5.5)$$

where each A_{kj} is either 0 or 1 depending on the correlation stencil used. In the simulation described here, we consider the nearest-neighbor stencil where each VCON output is correlated with its neighbors. Thus, $A_{kj} = 1$ for all $k \neq j$

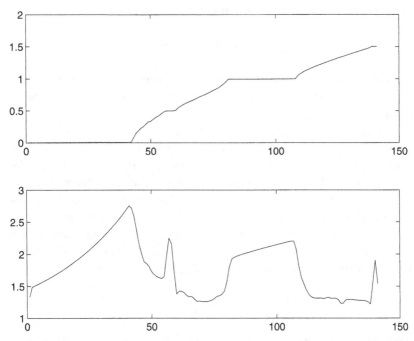

Figure 5.16. The top plot shows the activity (firing rate) of the input layer. The second plot shows the activity of the correlating layer.

for which $|k - j| \leq 1$. Figure 5.15 shows the circuit's wiring diagram and the output frequencies of all the VCONs.

We see that $f(t) = \cos_+ t/2$ causes the first layer to produce a stair-step function of frequencies, where plateaus indicate intervals of phase locking. This array is passed into the correlating layer, the outputs of which peak over the major intervals of phase locking from the first layer.

Various choices of $f(t)$ result in distinct outputs of the tonotope, up to the resolution of the network. We see in Figure 5.16 that the intervals of phase locking emerging from the gradient of VCONs are reflected in the peaks of frequency outputs of the correlating layer. This is essentially the power spectrum of output: At equilibrium firing frequencies, we have that

$$x_j \to \rho_j t + \phi_j.$$

Therefore, the correlating outputs eventually become

$$\lim_{t \to \infty} \frac{z_j(t)}{t} = \mu + \sum_{\rho_j m = \rho_k n} A_{k, j-N} \sum_{m,n=0}^{\infty} a_m a_n \cos_+ (m\phi_{j-N} - n\phi_k),$$

where $\{a_m\}$ are the Fourier coefficients of $\cos_+ x$:

$$\cos_+ x = \sum_{m=0}^{\infty} a_m \cos mx.$$

The differences of the phase deviations $\{\phi_m\}$ for phase-locked fibers appearing in this formula explains the variation in the output of the second layer in Figure 4.3.

Alternatively, the output can be read in another way. The neurotransmitter output of a neuron from a chemical synapse is proportional to the activity – or frequency of action potentials arriving at the synapse. Thus, a neural circuit like the tonotope would result in a profile of released neurotransmitter that describes the relative frequencies $x_N : \cdots : x_{2N-1}$; it breaks the input signal $f(t)$ down into a spatial pattern of chemicals that comprise the power spectrum of f.

It is known that Golgi cells of type II perform a correlating function like that described here [7], and the synchronization index observed in anesthetized cats is similar to that coming from the correlating layer.

5.8 Summary

This chapter introduces several interesting examples of small networks of neurons that are similar to ones occurring in biological systems. These models are simpler than the real biological systems they represent, and one must be careful in using their results without first checking a number of important factors. For example, the phenomenon of rhythm splitting is substantially more complicated than a bimodal VCON model described here.

The respiration model is quite simple, but the real system is still not completely understood. Still, the model provides some interesting guidelines for thinking about the breathing process. It does not describe how the energy needs of the organism influence the natural frequencies of the various neurons or of the diaphragm. Moreover, most animals have a complex system of pulmonary stretch receptors, but this model accounts for only one.

The area of biological rhythms is huge, and it continues to be one of the most actively researched in the life sciences. In addition, there is an emerging medical specialization in chrono-pharmacology and chrono-medicine, which focuses on when application of various drugs is most efficacious and when in a person's daily cycle of hormone baths is best to schedule surgery-recovery procedures. For general references on biological rhythms, see [142].

Finally, some issues of numerical algorithms for computer simulations were addressed. In these simulations, we sometimes used $V(x) = \cos x$ and $V_+(x) = \cos x \cos_+ x$ to finesse numerical accuracy and stability problems that can arise when $V(x) = \cos_+ x$ since its derivative is not continuous.

5.9 Exercises

1. *Respiration simulation.* Solve the respiration model numerically and present the solutions graphically by plotting a vertical mark at the firing of each neuron as well as the diaphragm displacement. Reproduce Figure 5.6.

2. *Diaphragm as a nonlinear spring.* Replace the term kz in the equation for z by the term $k(z + z^3)$ and carry out the simulation described in Figure 5.6 (see [56]).

3. *Rotation vector method applied to respiration.* Apply the rotation vector method to the respiration control model. In particular, evaluate the ratio $\phi(T)/(\omega_1 T)$ for large T and for the data used in Figure 5.6. Analyze the running respiration model as a function of forcing frequency ω. Determine the primary parameter regions of 1:1, 2:3, 3:5, 1:2, 2:5, and 1:3 phase locking of breath/stride. Note that when running, the diaphragm is stiffened, so k must be increased accordingly.

4. *Time crystal.* Relate the phase-response curves of Figure 5.13 to a time crystal as described in Figure 2.7.

5. *Numerical simulation of rhythm splitting.* Write a computer program that simulates the bimodal phase-locked loop. Reproduce Figure 5.13.

6. *Propagation time on an axon.* Estimate the propagation time of a signal in a circuit like that in Figure 5.14 having six VCONs. Simulate this numerically by stimulating the axon at one end with a pulse. Compare your predicted transit times with your computations.

7. *Binaural sound location.* Analyze the binaural model assuming that there are six receptor cells. Write a computer program to model the network. Treat each axon patch as being a VCON with $\omega = 0$. Estimate the propagation time of an excitatory signal from your computations. Using six receptor cells, identify parameters in the model that will most finely resolve the location of a noise.

6

Memory, phase change, and synchronization

In this chapter we study general networks of elements that are described by differential equations to see what information can be derived from them that is relevant to neurobiology. In the next chapter, we study some networks that have been mapped by neurobiologists, and we apply the results of this chapter to study those networks. The goal of both chapters is to describe temporal and spatial patterns of neuron firing in large-scale networks and to relate these results to controls and the flow, storage, and recall of information. Temporal patterns and the timing of signals is important in brain circuits [126, 18].

The first section presents and develops some relevant topics in network theory. Particularly important for our study are systems that are gradient-like. These systems have an associated scalar-valued function whose minima correspond to stable equilibrium values for the system. Knowledge of such functions facilitates not only the stability analysis of networks, but it also makes possible identification of critical parameters at which bifurcations occur.

We then study some aspects of VCON networks. Typically, there is an invariant torus of solutions, and the network can be reduced to consideration of phase variables alone. These in turn usually approach the form

$$\theta \to \omega t + \phi,$$

where ω is a vector of frequencies and ϕ is a vector of phase deviations.

The first step in analysis of output of such networks is to determine which elements have the same firing frequency and what are the differences in phase deviations between them. Through these two modes networks can carry and transmit information. Therefore, patterns of firing frequencies and of phase deviations are important.

The second section presents a theory of *mnemonic surfaces*. These surfaces arise through analysis of neural network models in which we can identify a potential function like those discussed in the preceding section. We refer to these as being mnemonic surfaces since they contain complete information about stable states of the network, and so they summarize concisely what the network has memorized. In addition to analysis of networks derived from neural network modeling, we show how to create a network with a prescribed mnemonic surface.

These mnemonic surfaces are not for the phase variables, but for the phase-deviation variables! This demonstrates, as we discussed in the preface, that synchronized neurons can communicate and their mode of communication is through timing. We can think of an FM radio where we tune a station by fixing a center frequency and then enjoy the signal that is carried by the phase deviations. Thus, the stable states of the network studied here have the form

$$\theta \approx \omega t + \phi,$$

where ω is the address of the channel of communication and ϕ deals with the timing and carries the message.

The third section investigates some aspects of signal processing in networks. This work develops the point made in the last paragraph. We develop in this section connections between neural networks and the rotation vector method that identifies an ensemble of frequencies and stable phase deviations during phase locking.

The fourth section presents the Fourier–Laplace method for determining emerging spatial patterns in a network. These are methods from calculus applicable to continuum approximations of the networks studied here, and they provide a methodology for detecting emerging patterns. For example, if the network elements are addressed in space by a continuous variable, say s, and if $\theta(s, t)$ denotes the phase variable of the element at site s at time t, then we might have that

$$\theta(s, t) \approx \omega(s)t + e^{p_k(s)t} \cos k \cdot s.$$

The spatial mode $\cos k \cdot s$ describes a spatial pattern. For example, if $p_k(s) \equiv 0$, then there is a spatial pattern of phase deviations.

The fifth section presents cellular automata that have been used to model and study networks. Many important aspects of neural networks can be described by formulating the original model directly in terms of matrices. This makes possible our use of powerful methods developed to study randomness using Markov's chains. These topics, and some interesting historical aspects are developed in this section.

Cellular automata describe neural networks by telling which are firing (on/off). These networks have played a significant role in many developments in mathematics (logic [96]), computers (Turing machines, [133]), and physiology (Rosenblueth, [137, 100]).

No attempt is made to describe all of the past work done on large discrete automata; in fact, only a few examples are presented here in detail. Among them are Hopfield's model, which is in the spirit of Ising models from physics, and Chetaev's models based on Markov chains.

The sixth section summarizes the results of this chapter and sets up the work in Chapter 7.

6.1 Network theory

Consider a network of elements that are each described by a collection of variables, say the jth one is described by a vector $\xi_j \in R^{n_j}$. The components of ξ_j might be chemical concentrations, reaction rates, ionic currents, membrane voltages, etc. The network we study is described by a system of differential equations of the form

$$\frac{dx_i}{dt} = F_i(x) \tag{6.1}$$

for $i = 1, \ldots, N$, where the (very large) vector x accounts for all of the individual descriptor vectors ξ_j, network connection strengths, etc., and F_i accounts for local dynamics of these variables as well as all of the network connections to it. Mathematical scientists and engineers have derived a great deal of information about such networks when certain assumptions are made about the elements and their interconnections.

6.1.1 Energy surfaces as descriptors of network dynamics

In this section, we first shape potential functions having prescribed stable states in rough analogy with the formation of memories in Freud's thought model. We then consider the associated dynamical system and the resulting description of an electrical process.

How can a smooth energy surface result from the description of a finite set of mechanical devices? The simplest case to consider is the simple harmonic oscillator

$$\ddot{u} + \omega^2 u = 0. \tag{6.2}$$

The solutions of this equation are $u(t) = A \cos(\omega t + \phi)$, where A and ϕ are some constants.

Multiplying both sides of the equation by $\dot{u}$ and integrating the result leads us to define the function $E(u, \dot{u})$ by

$$E(u, \dot{u}) = (\dot{u}^2 + \omega^2 u^2)/2.$$

E is called the energy of the oscillator, and the first term on the right is the kinetic energy and the second is the *potential energy*. Differentiating this function with respect to t gives

$$\frac{dE}{dt} = \nabla E \cdot (\dot{u}, \ddot{u})$$

$$= \omega^2 u \dot{u} - \omega^2 u \dot{u}$$

$$= 0.$$

Thus we see that the energy function is conserved along trajectories of the system, and the oscillator is referred to as being conservative.

Incidentally, solutions can be found by solving the equation

$$E = C,$$

where the constant C is the initial energy of the system. This is a first-order differential equation for u that can be solved by straightforward integration leading to the general solution given above.

The systems we deal with are not conservative; they are dissipative. For example, consider the damped harmonic oscillator

$$\ddot{u} + r\dot{u} + \omega^2 u = 0. \tag{6.3}$$

In this case, we have that

$$\frac{dE}{dt} = -r\dot{u}^2. \tag{6.4}$$

From this, we see that energy is removed from the system at a rate proportional to r, the damping coefficient. E will keep decreasing until $u = 0$ [58].

For visualization, it is appropriate to introduce a new variable: Let $\dot{v} = \omega u$. With this, the harmonic oscillator can be described by two first-order differential equations

$$\dot{u} = -ru - \omega v \tag{6.5}$$

and

$$\dot{v} = \omega u. \tag{6.6}$$

Now, $E = \frac{1}{2}(u^2 + \omega^2 v^2)$, and the surface $z = E(u, v)$ can be easily drawn. It is a paraboloid surface having minimum at $u = v = 0$. The above calculation shows that the point $(u, v, E(u, v))$ moves on this surface in such a way as to approach the minimum if $r > 0$. If $r = 0$, a solution moves in a circular orbit on the surface at a height determined by the system's initial energy.

Thus, a surface can be associated with a single oscillator, and its solutions can be depicted using the surface. Although there is but a single oscillator, there is a continuum of possible states for it, and the surface describes system responses. More important, it gives at once a description of the system's response everywhere! And therefore, a concise description of the system's solutions results.

6.1.2 Gradient systems

A number of interesting results have been found for networks when F is a *gradient function*. Such systems have appeared in other studies in biology and the neurosciences (see, e.g., [42], [48], and [129]). *Gradient* networks are those for which there is a scalar-valued potential function, say Φ, for which

$$F_i(x) = -\frac{\partial \Phi}{\partial x_i}(x)$$

for $i = 1, \ldots, N$. When this happens, we say that Equation (6.1) is a gradient system, and Φ is its potential function. The choice of the minus sign will be explained later.

A particularly nice feature of gradient systems is captured by the following calculation. Suppose that $x(t)$ is a solution of the general system

$$\dot{x} = -\nabla\Phi(x),$$

where $\nabla\Phi$ denotes the vector gradient of Φ:

$$\nabla\Phi(x) = \begin{pmatrix} \frac{\partial \Phi}{\partial x_1} \\ \vdots \\ \frac{\partial \Phi}{\partial x_N} \end{pmatrix}.$$

Differentiating $\Phi(x(t))$ along the solution gives

$$\dot{\Phi}(x) = \nabla\Phi \cdot \dot{x} = -\left|\nabla\Phi(x)\right|^2 \le 0.$$

This result shows that the value of Φ when it is evaluated along a solution of the system is always decreasing unless the solution is at a point where

$$\nabla\Phi(x) = 0.$$

Among such points are the candidates for extrema (minima, maxima, and saddle points) of Φ. The choice of using a minus sign here ensures that minima of Φ are stable points.

Being a gradient system imposes some conditions on F. In particular, for any $i, j = 1, \ldots, N$ we must have that

$$\frac{\partial F_i}{\partial x_j} = -\frac{\partial^2 \Phi}{\partial x_j \partial x_i} = -\frac{\partial^2 \Phi}{\partial x_i \partial x_j} = \frac{\partial F_j}{\partial x_i}$$

since the derivatives of Φ are assumed to be continuous. Thus, to be a gradient system, it is necessary that

$$\text{curl } F \equiv \nabla \times F(x) \equiv 0.$$

These are quite stringent restrictions on the system.

6.1.3 Gradient-like systems

Liapunov [93] pointed out that systems need not be gradient systems to behave like them. In particular, many important systems that are not gradient systems have associated with them a function that decreases along trajectories.

A useful example is

$$\dot{x} = Ax \tag{6.7}$$

where the eigenvalues of the symmetric part of the matrix A are all negative. Then the function $\mathcal{V}(u) \equiv u \cdot u$ is such a function:

$$\frac{d\mathcal{V}}{dt} = \dot{x} \cdot x + x \cdot \dot{x} \tag{6.8}$$

$$= (A + A^{tr})x \cdot x \tag{6.9}$$

$$\leq -2\lambda x \cdot x \tag{6.10}$$

$$= -2\lambda \mathcal{V}, \tag{6.11}$$

where $-\lambda$ is the eigenvalue of $A + A^{tr}$ nearest zero. Thus, $\mathcal{V}(t) \leq e^{-2\lambda t} \mathcal{V}(0)$.

More generally, one might consider a system

$$\dot{x} = F(t, x). \tag{6.12}$$

A function $\mathcal{V}(t, x)$ is called a Liapunov function for this system at an equilibrium, say $x = x^*$, if the following conditions are satisfied:

1. $F(t, x^*) = 0$ for all $t \geq 0$. (i.e., x^* is an equilibrium.)
2. $\mathcal{V}$ is a smooth function of t and x for $0 \leq t < \infty$ and for x near x^*.
3. $\mathcal{V}(t, x^*) < \mathcal{V}(t, x)$ for all values of x near x^*. ($\mathcal{V}$ has x^* as a local minimum.)
4. $\partial \mathcal{V}/\partial t + \nabla \mathcal{V} \cdot F(t, x) < 0$ for all t and for all x near x^*. ($\mathcal{V}$ decreases along solutions.)

We say that $\mathcal{V}(x)$ is a Liapunov function for Equation 6.1 if $\mathcal{V}$ is bounded from below and

$$\frac{d\mathcal{V}}{dt} \equiv \nabla\mathcal{V} \cdot F \le 0.$$

Any gradient system has a Liapunov function defined near a minimum of the potential function, namely the potential function itself! For example, if F is itself the gradient of a function Φ (so $F = \nabla\Phi$), then $\mathcal{V} = \Phi$ can be a Liapunov function for the system. In fact, in that case

$$\frac{d\mathcal{V}}{dt}(x) = \nabla\mathcal{V} \cdot \dot{x} = -|\nabla\mathcal{V}|^2.$$

Minima of a Liapunov function $\mathcal{V}$ correspond to stable equilibria for the system. Although such systems are still quite complicated, some useful facts can be derived for them.

Finding a Liapunov function is not easy [45]. Nevertheless, many interesting systems either are gradient systems, based on some known conserved quantity such as energy or momentum, or have another associated Liapunov function. There have been a number of attempts to relate memory in a neural network to potential wells of a potential function associated with a model neural network [33, 42].

An important consequence of these remarks is that systems having a Liapunov function behave similarly even when they are exposed to small amplitude, but highly irregular, noise.

Consider a dynamical system

$$\dot{x} = F(t, x)$$

that has a Liapunov function

$$\mathcal{V}(t, x)$$

near some equilibrium $x = x^*$. If this system is perturbed by noise, say

$$\dot{y} = F(t, y) + \varepsilon g(t, y),$$

where g is a bounded, but highly irregular, function that is at least Lebesgue integrable, then

$$\frac{d\mathcal{V}(t, y)}{dt} = \mathcal{V}_t + \nabla\mathcal{V} \cdot \big(F(t, y) + \varepsilon g(t, y)\big).$$

Thus, $\mathcal{V}$ is decreasing to some neighborhood of order $o(1)$ about x^* for small values of ε.

In rough terms, we can think of a parabolic bowl. If a marble is placed on the edge, it will eventually settle at the bottom point. Now, if the bowl is pummeled with small blows that leave shallow dents and if it is punctured with small holes, then the marble will still end up near the bottom, where "near" refers to the size of holes and dents. This is the idea of *stability under persistent disturbances*.

This idea is not the same as structural stability, where stability of an equilibrium must be preserved by perturbations. That is a more restrictive concept. The example

$$\dot{x} = -x^3 \tag{6.13}$$

is quite useful to keep in mind: $x = 0$ is stable under persistent disturbances, but it is not structurally stable [4].

6.1.4 Bifurcations and phase changes

Movement of solutions of Equation (6.1) is determined by the functions F_i. These in turn include numerous parameters that (ideally) are observable, or at least indirectly accessible, through experiments. These are denoted by vectors of parameters, say $\lambda_i \in R^{m_i}$, and we write

$$\frac{dx_i}{dt} = F_i(x, \lambda_i).$$

From the experimental side, the parameters describe variables that are constant (or almost constant) while the network elements (say, neurons) change on the t time scale, and λ_i parametrizes all possible biochemical and kinetic states of the ith element as well as connections within the network. From the mathematical side, these parameters describe the shape of the functions $\{F_i\}$.

The idea here is to use mathematical methods to derive from the general complex network simpler models that can be solved and whose solutions give some useful information about the full system. To do this, we study the dependence of solutions on parameters that appear in the system. These dependences may arise through scaling arguments, as in the study of *weakly connected networks*. In these cases, each network element is described by a separate differential equation, say

$$\frac{dx_i}{dt} = f_i(x_i)$$

for $i = 1, \ldots, N$, and the interconnections between separate elements are described by functions $\varepsilon g_i(x)$ telling what is the impact on site i of the entire network being in state x. The parameter ε measures the ratio of time scales between the element's response and the strength of coupling between network elements, so

$$\frac{dx_i}{dt} = f_i(x_i) + \varepsilon g_i(x).$$

In these cases, the general system has

$$F_i(x) = f_i(x_i) + \varepsilon g_i(x)$$

and advantage can be taken of the presence of the small parameter ε to study the system.

There is some evidence that neural networks are weakly coupled since a postsynaptic voltage change is approximately 1% the amplitude of an action potential that stimulated synaptic transmission [74]. We do not study these systems here, but refer the reader to [75]. Nor will we pursue perturbation methods further here. Instead, we consider general networks but only near bifurcation points.

Threshold, or bifurcation, phenomena occur in general networks at particular parameter or state values. Suppose that the system is operating in a state $x^*(t)$ when the system's parameters have the values $\lambda = \lambda^*$, where by λ we denote the collection of all relevant parameters in the system. We describe this by the notation

$$\frac{dx_i^*}{dt} = F_i(x^*, \lambda^*).$$

The linearization of the system about this state refers to the associated system

$$\frac{d\delta x_i}{dt} = A_i(x^*, \lambda^*)\delta x_i \qquad (6.14)$$

that is found by writing $x_i = x_i^* + \delta x_i$ in the general system, expanding in powers of δx_i, and ignoring higher order terms. Thus, the matrix A_i is defined by the partial derivatives of F_i:

$$A_i(x, \lambda) = \nabla_x F_i(x, \lambda).$$

The state $x^*(t)$ is *stable* if solutions of the linear system, Equation (6.14), are decaying to zero at an exponential rate [58]. However, the system can encounter a *bifurcation* if some of the eigenvalues of this problem meet the imaginary axis as system parameters change slowly, and it is at such points that solutions can experience a dramatic change in their behavior. For example, if the system is a gradient system and as parameters change a fold develops in the potential function, then a bifurcation occurs and usually the system can experience some sort of phase change.

As an illustration of this, consider the potential function

$$\Phi(x) = \lambda x_1 + \sin x_1 + \sum_{j=2}^{N} \frac{x_j^2}{2},$$

where λ is a scalar parameter. In this example, all of the dynamic variables

(x_i) are scalars and only one parameter appears. The gradient system for this potential is

$$\frac{dx_1}{dt} = -(\lambda + \cos x_1),$$

$$\frac{dx_i}{dt} = -x_i$$

for $i = 2, \ldots, N$.

The shape of Φ is simply a parabola in each of the variables $x_2, \ldots, x_N$, but the shape with respect to the first variable x_1 changes as the parameter λ changes. When λ is very large, the shape is essentially a straight line with a slight wiggle superimposed on it. When λ is near zero, the shape is essentially a sine function. As λ is decreased from large values, wrinkles develop in the line until at $\lambda = 1$ the wrinkles actually become dented, having negative slope at some places.

If $|\lambda| < 1$, there are two rest points in each interval of length 2π for the first equation. These are symmetrically distributed around π. One of them (the left one, denoted by $x_1^*(\lambda)$) is an exponentially stable rest point; the other one is a saddle point. However, when $|\lambda| > 1$, $x_1 \to \infty$ as $t \to \infty$. From the point of view of the rest points, the stable node and the saddle point come together and disappear as the bifurcation point is reached. Also, the appropriate linearization for the system near the stable rest point is

$$\frac{d\delta x_1}{dt} = \sin x_1^*(\lambda)\delta x_1,$$

$$\frac{d\delta x_i}{dt} = -\delta x_i$$

for $i = 2, \ldots, N$, and at $\lambda = 1$, $x_1^*(1) = \pi$ at which the right-hand side of the first equation is 0.

Thus, we have seen three ways to view the phenomenon of bifurcation:

- through the geometrical structure of the potential function,
- through the existence and stability of rest points for the system, and
- through the algebraic structure of an associated linear system's eigenvalues.

We will use all three approaches in studying bifurcations in neural networks.

One final point about this example: As $|\lambda|$ approaches 1, $x_1^* \to \pi$. In this case the analogue of Equation (6.14) is

$$\delta\dot{x}_1 = \sin x_1^*(\lambda)\delta x_1,$$

and we see that the eigenvalue of the system (namely, $\sin x_1^*(\lambda)$) hits zero when

$|\lambda|$ hits 1. Note that when $|\lambda| = 1$, small constant input can have a dramatic influence on solutions

$$\dot{x}_1 = 1 + \varepsilon\alpha + \cos x_1.$$

If $\alpha < 0$, then x_1 stabilizes; if $\alpha > 0$, then $x_1 \to \infty$ no matter how small is ε. In this last case, the solution moves very slowly ($\dot{x}_1 \approx 0$) when $\cos x_1 \approx -1$. Thus, if the bifurcation point is approached from above, the oscillation ($\cos x_1$) maintains constant amplitude but its frequency approaches zero.

Since there is a great deal of redundancy in neural networks, bifurcations in them usually involve many network elements (almost) simultaneously. Therefore, we must cope with multiple bifurcations. Fortunately, the redundancy of the system usually makes it possible to uncouple the system near bifurcations.

It is through a study of bifurcations that we can investigate phase changes in a network; these are situations where slight changes in initial conditions or in system parameters can result in dramatic changes in the network's behavior. Near bifurcations, it is often possible to greatly simplify the problem and so derive a tractable mathematical model, called a canonical model. Various bifurcations are characterized by their associated *canonical problem* [130].

Bifurcations have been studied usefully in a variety of applications. Determining thresholds of change in a system's behavior is important in many applications in the physical and life sciences, for example, in examining

- population size thresholds for propagation of an epidemic [86],
- explosion limits for combustion reactions [118],
- quenching and propagating chain-branched chemical reactions [37], and
- magnetization of ferromagnets as a function of temperature [101].

These studies show that bifurcations can be important guides to experiments since qualitative changes in a system's behavior are often observable in experiments, and these experiments in turn can serve to help researchers identify model parameters. These phenomena are sometimes referred to as being phase changes, in analogy with bifurcations that occur as water passes from a gas to a liquid to a solid as temperature is decreased. Bifurcations prove useful in our studies of neural networks as well.

6.1.5 VCON networks

The example in the preceding section suggests that VCON models arise naturally as canonical models near saddle-node bifurcations. This is indeed the case as shown, for example, in [23]. VCON networks do arise in a natural way as the canonical model for a saddle-node bifurcation on or near a limit cycle, and so studying them is not only a study of a peculiar circuit derived from abstraction of neural activity, but it is a study of *any* system near a multiple saddle-node bifurcation. These are among the more difficult (elementary) bifurcations to study since global information is needed to follow solutions after encountering one. The VCON networks go beyond the canonical equations, which are valid only for the state variable near the bifurcation value, because they provide global information. They characterize a saddle-node bifurcation on a limit cycle.

A general VCON network (ignoring synaptic delays) is described by the equations

$$\dot{\theta}_j = a_j, \tag{6.15}$$

$$\tau_j \dot{a}_j + a_j = P_j(a_j, \theta_j) \tag{6.16}$$

for $j = 1, \ldots, N$. In all of these models we assume that

- $a_j \in E^{m_j}$;
- $\theta_j \in T^{m_j}$, by which we denote the m_j-dimensional torus;
- the variables in $\theta = (\theta_1, \ldots, \theta_N)$ denote firing phases;
- those in a denote activities (here they are firing rates);
- those in P_j describe local dynamics and network connections; and
- the numbers $\tau_j \geq 0$ give time scale ratios

We show in [75] that the complexity of this network can be reduced to consideration of phase variables alone. The idea is to find a function $a = A(\theta)$ that is invariant under the flow of the system. The surface described by this equation is a torus T^N. We do not pursue this further in the text, but instead restrict attention only to behavior of the phase variables:

$$\dot{\theta}_j = A_j(\theta).$$

6.2 Mnemonic surfaces

Ideas starting with von Helmholtz [48], and carried on by Bruecke, Freud, and Jung [77], revolved around attempts to find electrochemical bases for human psychology. In particular, they sought an energy function associated with neural activity in analogy with the physics of mechanical systems. Such an energy function was sometimes referred to as being psychic energy. Freud's thinking about this is evidenced by the following quote:

I therefore postulate that for the sake of efficiency the second system succeeds in retaining the major part of its cathexes of energy in a state of quiescence and in employing only a small part of displacement. The mechanics of these processes are quite unknown to me; anyone who wished to take these ideas seriously would have to look for physical analogies to them and find a means of picturing the movements that accompany excitation of neurones [33, p. 638].

With circuit analogs of neural networks, we are presented the opportunity to picture such movements. In fact, the VCON model has a mechanical analogue in coupled systems of pendulums, where energy can be defined using routine physical principles. The abstraction presented by the model is a dynamical system for which there is a potential function for phase deviations under various natural conditions analogous to energy.

What follows in this section is a derivation of energy surfaces and dynamical systems derived from them. We will return in a later section to the VCON model, and consider it from the point of view of mnemonic surfaces – surfaces that represent memories. In this section, we develop a theory of mnemonic surfaces and derive a variety of interesting simulations using them.

The word *mneme* was used often by Freud; its definition is from mnemon (Greek for mindful), and it is used now more frequently as mnemonic: of or pertaining to memory.

6.2.1 Gaussian well construction of mnemonic surfaces

We can construct some interesting mnemonic surfaces to serve as potential functions for model neural networks.

Visualize a slab of dripless candle wax. When heat is applied at one point on it, a *dimple*, or *Gaussian well*, will melt in it. The longer the stimulus is applied, the broader and deeper will be the well. The model described here captures a similar process creating a gradient-like system.

We consider a network of elements (e.g., circuits) whose overall state is described by a state vector $u = (u_1, u_2, u_3, \ldots, u_N)$. These variables might be frequencies, voltages, etc. The index of each state variable may be related to the position of a network element, but not necessarily. This is discussed later.

The following model accepts an input state vector $u = y^*$, and it creates a basin of attraction for y^* whose breadth and depth increase with the length of time the input is applied. In a sense, the stimulus causes an energy surface to dimple or be dented. After many stimuli have been applied, the surface will be irregularly dented. When the system is reinitialized with a new value for $u(0)$, say reflecting a new focus of attention, the system will schuss to the bottom of a dent. Note that attraction to a basin is based on similarity of state variables, for example, correlation, and not on spatial proximity.

We consider the quasi-gradient system

$$\dot{u} = -\nabla_u \Phi\big(y(\varepsilon t), u\big), \qquad (6.17)$$

where y denotes the history of stimuli applied to the network and ε denotes the ratio of the learning time scale to the system response ($\varepsilon \ll 1$, so the system is gradient-like on the t scale).

The stimulus history, or training sequence, y is a jump function listing the stimulus state vector, the length of time it is applied, and its amplitude. At time t, there have been applied a number $M(\varepsilon t)$ of different stimulus values, say $y_0, y_1, \ldots, y_M$, with corresponding application times $t_1, t_2, \ldots, t_M$. Therefore, the time interval is divided into $[0, \varepsilon t) = [0, \varepsilon t_1) + [\varepsilon t_1, \varepsilon t_2) + \cdots + [\varepsilon t_M, \varepsilon t)$ over which y takes its constant values, respectively.

The function Φ is defined in the following way. Let dimple functions be defined over these intervals:

$$\Phi_j(\varepsilon t, u) = -\varepsilon(t - t_j) A_j \exp\left(-\frac{|u - y_j|^2}{2\varepsilon(t - t_j)}\right) \qquad (6.18)$$

for $t_j \leq t < t_{j+1}$, $\Phi_j(\varepsilon t, u) = \Phi_j(\varepsilon t_{j+1}, u)$ for $t > t_{j+1}$, and $\Phi_j(\varepsilon t, u) = 0$ for $t < t_j$. Finally, let

$$\Phi\big(y(\varepsilon t), u\big) = \sum_{j=0}^{M(\varepsilon t)} \Phi_j(\varepsilon t, u), \qquad (6.19)$$

where $t_{M(\varepsilon t)}$ is the last switching time before t.

This model is concocted so that the width of a dimple is like the variance of a Gaussian distribution. Thus, the input history takes the form $\{A_j, y_j, t_j, \tau_j \equiv t_{j+1} - t_j, \varepsilon\}$, which gives the stimulus strength and value, the switching time, the stimulus duration, and the learning time scale.

6.2.1.1 Example: a single dimple

Consider a single state variable $u(t)$. We apply a stimulus of strength A at $y = 0.0$, and we leave it on. Then

$$\Phi\big(y(\varepsilon t), u\big) = -\varepsilon t A \exp\left(-\frac{u^2}{2\varepsilon t}\right). \qquad (6.20)$$

As a result the equation for $u(t)$ becomes

$$\dot{u} = -u A \exp\left(-\frac{u^2}{2\varepsilon t}\right). \qquad (6.21)$$

u eventually approaches 0.0 like $\exp(-At)$. ($V(u) = u^2/2$ is a Liapunov function for this equation.)

6.2.1.2 *Example: two dimples*

Consider the preceding case but with two stimuli, one at $y = 0$ and one at $y = y_2$, each having been applied for equal lengths of time, say τ:

$$\dot{u} = -u \exp\left(-\frac{u}{2\varepsilon\tau}\right) - (u - y_2)\exp\left(-\frac{|u - y_2|^2}{2\varepsilon\tau}\right). \qquad (6.22)$$

In this case, Φ has two minima, one near 0 and the other near y_2. This example shows that the original attractor is distorted by the second input. If $y_2/\sqrt{\varepsilon\tau}$ is far from 0, then the second term on the right is negligible for u near zero. If it is near 0, then there is significant distortion. Note also that there is now an unstable equilibrium somewhere between 0 and y_2.

6.2.1.3 *Example: a mature surface*

Suppose that various stimuli have been applied up to time t_M, but none thereafter. Then for $t > t_M$, we have

$$\dot{u} = -\sum_{j=0}^{M}(u - y_j)A_j \exp\left(-\frac{|u - y_j|^2}{2\varepsilon\tau_j}\right). \qquad (6.23)$$

It is clear that if the values y_j are widely separated and if $u(0)$ is near some y_J, then x approaches near y_J as $t \to \infty$. However, the actual surface Φ over E^N is quite complicated.

The model presented here memorizes the data $y_0, y_1, \ldots, y_M$. It is a gradient-like system that can be shaped, or sculpted, by inputs to it. The time scale ε is introduced to ensure that the responses take place before new stimuli become dominant. One expects that if a stimulus is applied at y^* for a length of time, then any (not too distant) initial point will evolve into y^*. If a new stimulus is applied at another point, say y_1, but everything else is kept the same, the solution will remain at y^*. An example of this is shown in Figure 6.1.

6.3 Signal processing: frequency and phase deviations

In outline, our study of signal processing in VCON networks proceeds in the following way:

- The general network model we consider is in the activity-phase form

$$\dot{\theta} = a,$$

$$\tau\dot{a} + a = \mathcal{P}(a, \theta).$$

- We determine an invariant torus of solutions

$$a = A(\theta).$$

Mnemonic Surface

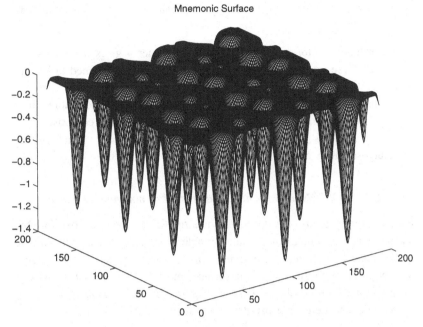

Figure 6.1. A mature surface of Gaussian wells.

- On this torus the problem takes the form

$$\dot{\theta} = \omega + F(\theta).$$

- We seek a solution of the form

$$\theta = \Omega t + \phi,$$

where the vector of frequencies (Ω) and the vector of phase deviations (ϕ) are to be determined.

- There results a system for the phase deviations ϕ among the oscillators that are frequency locked:

$$\dot{\phi} = G(\tilde{\phi}) + h.f.t..$$

Here $\tilde{\phi}$ is a truncation of ϕ that accounts for phase deviations among those oscillators that are synchronized and $h.f.t.$ refers to higher-frequency terms.

- If the system

$$\dot{\phi} = G(\tilde{\phi})$$

obtained from the phase-deviation equation by ignoring higher-frequency terms stabilizes $\tilde{\phi}$, then the phase deviations have been captured, and the network is operating in a stable way. This can happen if the function G is gradient-like.

The rotation vector method described in Chapter 4 gives conditions under which these steps can be successfully carried out and applied to VCON networks.

In this section we still consider general networks, but we add more structure to them. In particular, we assume that the state vector x comprises both amplitude variables (referred to here as activities a) and phase variables θ that describe oscillatory aspects of network elements.

6.3.1 A VCON network having a prescribed mnemonic surface

For the moment let us consider u_1 in Equation (6.17) to be a timelike variable and focus attention on the remaining variables $u_2, \ldots, u_N$. Given a mnemonic surface described by $\Phi(u)$, where $u = (u_2, \ldots, u_N)$, we can easily design a VCON circuit with this mnemonic surface embedded in it in such a way that the phase deviations of the network are captured by energy wells in Φ.

With this, we define a phase-detector system by the formulas

$$P_j = \frac{\partial \Phi}{\partial \theta_j}.$$

A pivotal question is how the connection matrices of the underlying network must be constructed to ensure such a form for the inputs. The connection structure underlying this network can be determined by calculating the sensitivity matrix

$$\frac{\partial \dot{\theta}_j}{\partial \theta_i}.$$

Essentially, this is determined by the matrix

$$\nabla \nabla^{tr} \Phi$$

This process is illustrated next.

6.3.1.1 Φ embedded in a VCON network

We can embed the mnemonic surface in a VCON network in the following way: Let us consider a network of N model neurons having phase vector θ. We will take the driving frequency vector to be

$$\omega = \begin{pmatrix} 1 \\ \vdots \\ 1 \end{pmatrix}.$$

Then we define a sequence of vectors $W_2, \ldots, W_N$ to span E^{N-1} and be orthogonal to ω as in Section 4.5, or, W can be defined by the formulas

$$
W_2 = \begin{pmatrix} 1 \\ -1 \\ 0 \\ \vdots \\ 0 \end{pmatrix}, \ldots, W_N = \begin{pmatrix} 0 \\ 0 \\ \vdots \\ 1 \\ -1 \end{pmatrix}.
$$

Then, we consider the network

$$
\dot{\theta} = \omega + W \nabla_u \Phi(A\theta), \tag{6.24}
$$

where the matrix A is defined by

$$
A = (W^{tr} W)^{-1} W^{tr}.
$$

The relation of this to the VCON model is that

- $\tau = 0$,
- $P = W \nabla_u \Phi(A\theta)$,
- the synapses are fast, and
- there are no chemical synapses.

Setting $\theta = \omega v + W u$, we have

$$
\dot{v} = 1, \tag{6.25}
$$

$$
\dot{u} = \nabla \Phi(u). \tag{6.26}
$$

As before, we see that if $u \to u^*$, then

$$
\theta \to \omega t + W u^*
$$

as $t \to \infty$.

The connections prescribed by this network are determined by calculating the sensitivity matrix as described above. We have

$$
\frac{\partial \dot{\theta}_j}{\partial \theta_i} = \sum_{k=2}^{N} W_{jk} \sum_{m=2}^{N} \frac{\partial^2 \Phi}{\partial u_m \partial u_k} A_{mi}.
$$

Thus, the connection matrix when the system has arrived at the equilibrium u^* of Φ is

$$
C = W \frac{\partial^2 \Phi}{\partial u^2}(u^*)(W^{tr} W)^{-1} W^{tr}.
$$

Note that this matrix C is symmetric, indicating that the mnemonic surface is restricted to describing networks having symmetric connections.

6.3.2 Surfing a quasi-static mnemonic surface

Finally, we come to the dynamics described by such mnemonic surface models.

We suppose that initial conditions are presented to the network from outside, perhaps by the searchlight mechanism described later where a more persistent (or important) signal will reinitialize the network. Once initialized, the network begins to function according to the results derived in the preceding sections. The trajectory will move in such a way as to approach a minimum of Φ.

Obviously, the value of surfaces for visualization decreases when the number of neurons exceeds 2! However, the fact that there is a gradient system describing the system's dynamics greatly simplifies analysis of it. Many analytic techniques are available for studying such systems, e.g., see [123].

In rough terms, we suppose the surface to be describing the dynamics of the neocortex. Then an input from the thalamus-reticular complex network will excite activity at some points of the neocortex. This initializes the gradient system, and the system proceeds to move signals in such a way as to reach a minimum of the potential function. A new stimulus will reinitialize the network, and the process continues.

Thus, we can visualize the system's dynamics in terms of a surfer who is placed on the sculpted potential surface and moves toward a minimum. A new stimulus corresponds to picking the surfer up and placing him at another point on the potential surface.

The surface is sculpted by formation during developmental periods, including genetic coding, and by learning through experiences from that point on. For example, Φ might have the form

$$\Phi(u_2, \ldots, u_N) = \sum_{m=1}^{M} A_m \exp\left[\frac{\sin^2\left(\sum_{j=2}^{N}\left(u_j - c_{j,m}^0\right)\right)}{\sigma_m} \right], \qquad (6.27)$$

where $\vec{c}_m^0 = (c_{2,m}^0, \ldots, c_{N,m}^0)$ for $m = 1, \ldots, M$ are centers of the Gaussian wells. Such functions can be constructed as shown in the Section 6.2.1. A solution of the model will toward a minimum of Φ. Minima can change slowly, and the system's solution will track one it is near. If the system is reinitialized by external changes, then the solution emanating from the new initial position will surf to a nearby minimum.

6.4 Fourier–Laplace methods for large networks

Although networks of neurons are made up of discrete elements, it is useful to approximate a discrete array of them by a continuum, thereby enabling us to use methods of calculus.

We consider now a continuous array of circuit elements. Let $s \in R^3$ be a continuous variable that locates sites (addresses) at which are circuit elements, let $\theta(s, t)$ denote the phase of the circuit element at position s, and consider the dynamic equation

$$\tau(s)\frac{\partial^2 \theta}{\partial t^2} + \frac{\partial \theta}{\partial t} = \omega(s) + S\left\{V\left[\theta(s, t)\right] + e(s, t)\right\},$$

where $e(s, t)$ describes the total input from other sites and S is a sigmoidal function as described earlier. We suppose that

$$e(s, t) = \int_{-\infty}^{\infty} \int_{-\infty}^{\infty} \int_{-\infty}^{\infty} \tilde{K}(s, s')V\left[\theta(s', t)\right] ds'.$$

The function $\tilde{K}$ is called the *kernel* of the integral, and it describes the strength of connection from site s' to site s. If there is an excitatory connection from site s' to site s, then $\tilde{K}(s', s) > 0$; if it is inhibitory, then $\tilde{K}(s', s) < 0$. The strength of the connection in either case is described by $|\tilde{K}(s', s)|$. Thus, the connections are described by the polarity (sign) of $\tilde{K}$ and its amplitude.

6.4.1 Frequency-domain analysis of the continuum model

Of primary concern for us in this investigation is the determination of firing distributions of the form

$$\theta(s, t) \approx \Omega(s)t + \Phi(s, t),$$

where $\Phi(s, t)/t \to 0$ as $t \to \infty$. This entails finding the distribution of firing frequencies $\Omega(s)$ and the quasi-static phase deviation Φ [58]. In the next chapter, we will apply these methods to study some specific brain circuits.

Because of the low-pass filter, we only consider terms on the right that are essentially constant, so we replace

$$S\left\{V\left[\theta(s, t)\right] + e(s, t)\right\}$$

by

$$S_0 + \int_{-\infty}^{\infty} \int_{-\infty}^{\infty} \int_{-\infty}^{\infty} K(s, s')f\left[\theta(s, t) - \theta(s', t) + \psi(s, s')\right] ds',$$

where $f(0) = 0$ and S_0 is a constant. The result is

$$\tau(s)\frac{\partial^2 \theta}{\partial t^2} + \frac{\partial \theta}{\partial t} = S_0 + \omega(s) + \int_{-\infty}^{\infty} \int_{-\infty}^{\infty} \int_{-\infty}^{\infty} K(s, s')$$
$$\times f\left[\theta(s, t) - \theta(s', t) + \psi(s, s')\right] ds'.$$

Substituting $x = \Omega t + \Phi$ in this equation gives

$$\tau(s)\frac{\partial^2 \Phi}{\partial t^2} + \frac{\partial \Phi}{\partial t} = S_0 + \omega(s) - \Omega(s) + \int_{-\infty}^{\infty} \int_{-\infty}^{\infty} \int_{-\infty}^{\infty} K(s, s')$$

$$\times f\left\{ [\Omega(s) - \Omega(s')]t + \Phi(s, t) - \Phi(s', t) \right\} ds',$$

and we must determine both the frequency Ω and the phase deviations $\Phi(s, t)$ from this single equation.

We suppose that:

- Φ changes slowly with respect to t. We write $T = \varepsilon t$ and

$$\Phi = \Phi(s, T)$$

  to emphasize this slow variation of Φ.

- $\omega(s) \equiv \omega_0$ is a constant.

With these assumptions, we take

$$\Omega(s) \equiv S_0,$$

and the model becomes

$$\tau \varepsilon^2 \frac{\partial^2 \Phi}{\partial T^2} + \varepsilon \frac{\partial \Phi}{\partial T} = \omega_0 + \int_{-\infty}^{\infty} \int_{-\infty}^{\infty} \int_{-\infty}^{\infty} K(s, s') f\left[\Phi(s, T) - \Phi(s', T) \right] ds'.$$

We obtain a self-consistent solution to this problem that depends only on s (not t) if there is a function $\Phi(s, t) = \Phi_0(s)$ that makes the right-hand side zero. To investigate the possibility of such solutions, we consider the equation with $f(u) = \cos \alpha u = \mathbb{R}\, e^{i\alpha u}$, where $\mathbb{R}\, v =$ the real part of v:

$$-\omega_0 = \int_{-\infty}^{\infty} \int_{-\infty}^{\infty} \int_{-\infty}^{\infty} K(s, s') f\left[\Phi_0(s) - \Phi_0(s') \right] ds' \qquad (6.28)$$

$$= \int_{-\infty}^{\infty} \int_{-\infty}^{\infty} \int_{-\infty}^{\infty} K(s, s') \cos\left\{ \alpha\left[\Phi_0(s) - \Phi_0(s') \right] \right\} ds' \qquad (6.29)$$

$$= \mathbb{R} \int_{-\infty}^{\infty} \int_{-\infty}^{\infty} \int_{-\infty}^{\infty} K(s, s') \exp i\alpha\left[\Phi_0(s) - \Phi_0(s') \right] ds'. \qquad (6.30)$$

We expect that a solution Φ_0 gives a useful approximation to Φ (i.e., $\Phi = \Phi_0(s) + O(\varepsilon)$, as in [58]).

Let us consider the associated problem for Equation (6.28) obtained by ignoring the real part operation $\mathbb{R}$:

$$-\omega \exp\left[-i\alpha \Phi_0(s) \right] = \int_{-\infty}^{\infty} \int_{-\infty}^{\infty} \int_{-\infty}^{\infty} K(s, s') \exp\left[-i\alpha \Phi_0(s') \right] ds'.$$

This is a familiar equation: It is a Fredholm equation of the second kind, and it can be solved in a variety of ways depending on the properties that the kernel K has.

Details at this level of analysis move beyond the scope of the present book, but a typical analysis proceeds as follows: Most kernels K can be approximated by sums of the form

$$K(s', s) = \sum_{j=1}^{L} f_j(s)g_j(s'),$$

where f_j and g_j are some appropriate independent functions. The matrix

$$M_{j,k} = \iint f_j(s)g_k(s)\,ds$$

arises in the analysis of this equation, and if $-\omega_0$ is an eigenvalue of M, then there is a solution of the Fredholm equation for $\exp[-i\alpha\Phi_0(s)]$. The analysis is completed by passing to the limit $L \to \infty$.

Let us write one such solution as $\Phi_0(s)$. Then the solution of the original problem will be approximately

$$\theta(s, t) \approx (S_0 + \omega_0)t + \Phi_0(s),$$

which establishes the existence of spatial patterns of phase deviations. Relations between these solutions and steady progressing waves are developed in Chapter 7.

If connections are distributed according to a normal probability distribution, then it is appropriate to take for K a normal distribution of influence from any site. A *normal distribution* is described by the function

$$\mathcal{N}(s) = e^{-s^2/2\sigma^2}/\sqrt{2\pi\sigma^2},$$

and the probability of finding a connection within a distance s^* away from $s = 0$ is given by the integral

$$\int_{-s^*}^{s^*} \mathcal{N}(s)\,ds.$$

Approximately 68.3% of the connections are within σ units, 95.4% are within 2σ, and 99.7% are within 3σ of $s = 0$. Combinations of such functions can be used to approximate many reasonable influence functions.

Gaussian functions, or normal distributions, arise naturally in models. It is known from the Central Limit Theorem of probabilty theory that most random processes, when correctly scaled, have (approximatly) normal distributions. Gaussians also have the nice property that their Fourier transform is easy to calculate:

$$\tilde{\mathcal{N}}(k) = \int_{-\infty}^{\infty} e^{-iks}\mathcal{N}(s)\,ds$$

$$= e^{-k^2\sigma^2/2}.$$

The inverse transform gives

$$\mathcal{N}(s) = \frac{1}{2\pi}\int_{-\infty}^{\infty} e^{iks}\tilde{\mathcal{N}}(k)\,dk.$$

Another useful kernel is the *characteristic function* of an interval $(-L, L)$:

$$K(s) \equiv \frac{1}{2L}\mathbf{1}_{-L,L}(s),$$

which has the value $1/2L$ when $s \in (-L, L)$ and is zero otherwise. This function indicates that there are connections of equal strength out a distance L from $s = 0$ and none beyond that. The Fourier transform of the characteristic function is

$$\tilde{K}(k) = \text{sinc}(kL) \equiv \frac{\sin kL}{kL}.$$

We will investigate this kernel in the next chapter.

Wavelets [125] also provide interesting kernels. These are denoted by

$$K(s) = W(s, L).$$

In each case L measures the spread of influence of the kernel.

6.4.2 *One-dimensional patterns*

In addition to the steady state distributions for Φ that we found in the preceding section, we can study the temporal emergence of patterns in networks like these. First, we linearize the system about a known or prospective state; then we analyze the stability of various modes by taking the Laplace transform in time and the Fourier transform in space. If one of the modes grows faster than the others, then that mode will be asserted as an emerging pattern which will have the wave number of that mode.

Let us consider the model from the preceding section, but with $s \in R^1$. To illustrate the basic computations involved in finding patterns, we consider a simple kernel K where connections have the same polarity and amplitude over a fixed interval:

$$K(s) = \frac{1}{2L}\mathbf{1}_{-L,L}(s) \equiv 1 \quad \text{if} \quad |s| < L$$

and zero otherwise.

We will also use the Laplace transform of functions. This is defined for a function $g(t)$ by

$$\hat{g}(p) = \int_0^\infty e^{-pt} g(t)\, dt.$$

The inverse Laplace transform is given by

$$g(t) = \frac{1}{2\pi i} \int_{\sigma - i\infty}^{\sigma + i\infty} e^{pt} \hat{g}(p)\, dp,$$

where the line $\mathbb{R}\, p = \sigma$ is chosen to lie to the right of all singularities of $\hat{g}$. Laplace transforms are discussed in Appendix A.

Suppose that $\Phi = \Phi^*$ is a uniform (constant) solution for the equation

$$\tau \varepsilon^2 \frac{\partial^2 \Phi}{\partial T^2} + \varepsilon \frac{\partial \Phi}{\partial T} = \int_{-\infty}^\infty K(s - s') f\big[\Phi(s, T) - \Phi(s', T)\big]\, ds'.$$

We consider perturbations of this state and try to determine if any of them can grow. We write $\Phi = \Phi^* + \phi$. Then

$$\tau \varepsilon^2 \frac{\partial^2 \phi}{\partial T^2} + \varepsilon \frac{\partial \phi}{\partial T} = \int_{-\infty}^\infty K(s - s') f\big[\phi(s, T) - \phi(s', T)\big]\, ds'.$$

Linearizing the nonlinear term in the integrand about $\phi = 0$ gives

$$\tau \varepsilon^2 \frac{\partial^2 \phi}{\partial T^2} + \varepsilon \frac{\partial \phi}{\partial T} = \int_{-\infty}^\infty K(s - s') f'(0)\big[\phi(s, T) - \phi(s', T)\big]\, ds'.$$

The idea here is to study the linear problem to determine if there are some combinations of spatial modes, which we determine using Fourier transforms, and dominant temporal growth, which we determine using Laplace transforms, that describe possible emerging patterns of growth.

Taking the Fourier transform in s and the Laplace transform in t of both sides gives an equation relating the amplification rate (p) with the wave number (k). On the left side of the equation, we get

$$\tau \varepsilon^2 p^2 \breve{\phi} + \varepsilon p \breve{\phi}$$

from direct application of the Laplace transform. On the right-hand side, the first term is

$$\int_{-\infty}^\infty K(s - s') f'(0)\, ds'\, \phi(s, t) = \int_{-\infty}^\infty K(u)\, du\, f'(0) \phi(s, t)$$

$$= \tilde{K}(0) f'(0) \phi(s, t),$$

and the second term is a convolution integral whose Fourier transform is the product of the transforms. Thus, the Fourier transform of the right-hand side is

$$\left[\tilde{K}(0)f'(0) - \tilde{K}(k)f'(0)\right]\check{\phi}\ (k, p) = \left[1 - \text{sinc}(kL)\right]f'(0)\check{\phi}\ (k, p).$$

Therefore, the *dispersion relation* is

$$\tau\varepsilon^2 p^2\check{\phi} + \varepsilon p\check{\phi} = \left[1 - \text{sinc}(kL)\right]f'(0)\check{\phi}.$$

There is a value of k for which p is a maximum, and it is that wave number that will eventually dominate the solution of the linear problem. That value, say $k = k^*$, occurs the first time when $\tan kL = kL$. The solution of the linear problem will then eventually have the form

$$\phi(s, t) \approx \exp\left[p(k^*)t\right]\cos(k^*s + \psi)$$

for some constant ψ, and the emerging pattern will have wave number k^*.

Similar computations can be carried out for normal distributions of connections.

6.4.3 More complicated connection kernels

If there is a single Gaussian distribution of positive (excitatory) connections, then the stability analysis described in the last section results in nothing much happening: All phases tend to the equilibrium value Φ^* since the Fourier transform of the kernel is always positive.

What happens if there are negative feedbacks in the network? To answer this, we consider an example in which there are normal distributions of excitatory and inhibitory connections having different variances. For example,

$$K(s) = \sum_j g_j \frac{\exp\left[\frac{-s^2}{2\sigma_j^2}\right]}{\sqrt{(2\pi\sigma_j^2)}},$$

where the weights $g_1, g_2, \ldots$ need not all be positive. In fact, let us consider the special case where

$$K(s) = \frac{g_1}{\sqrt{2\pi\sigma_1^2}}\exp\left[\frac{-s^2}{2\sigma_1^2}\right] - \frac{g_2}{\sqrt{2\pi\sigma_2^2}}\exp\left[\frac{-s^2}{2\sigma_2^2}\right] + \frac{g_3}{\sqrt{2\pi\sigma_3^2}}\exp\left[\frac{-s^2}{2\sigma_3^2}\right].$$

Figure 6.2 shows the case where

$$K(s) = \frac{3}{\sqrt{2\pi}}\exp\left[\frac{-s^2}{2}\right] - \frac{5}{\sqrt{4\pi}}\exp\left[\frac{-s^2}{4}\right] + \frac{2}{\sqrt{6\pi}}\exp\left[\frac{-s^2}{6}\right].$$

Both the kernel and the dispersion function $p(k) \propto \tilde{K}(0) - \tilde{K}(k)$ are shown. These plots indicate that there is a most unstable mode for growth of patterns.

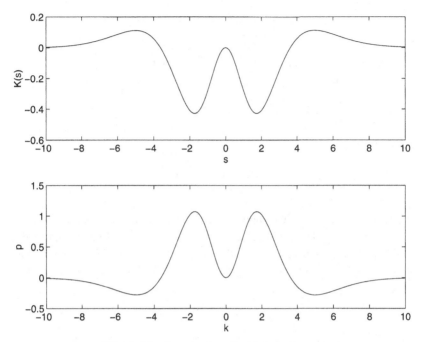

Figure 6.2. Gaussian kernel with three terms.

We proceed with the linear stability analysis of

$$\tau\varepsilon^2\frac{\partial^2\Phi}{\partial T^2} + \varepsilon\frac{\partial\Phi}{\partial T} = \int_{-\infty}^{\infty} K(s-s')f\big[\Phi(s,T) - \Phi(s',T)\big]\,ds'$$

with this choice of kernel. We start with the dispersion relation

$$\tau\varepsilon^2 p^2 + \varepsilon p = f'(0)\big[\tilde{K}(0) - \tilde{K}(k)\big].$$

Figure 6.2 shows p versus k for $\tau \approx 0$ and $f'(0) = 1$. Note that there is a value k at which p is maximum. This is called the *most unstable mode*. When the system evolves from x^*, the pattern that will be observed will have wave number $k_c \approx \pm 4.17$. That is, it will be of the form $B\cos k_c s + C\sin k_c s$. If we were able to observe this system, then we would see an action potential pattern emerge that has discernible spatial structure:

$$\Phi(s,t) \approx \exp\big[\varepsilon p(k_c)t\big](B\cos k_c s + C\sin k_c s).$$

The other modes of Φ are (relatively) damped.

6.4.4 Two-dimensional patterns

Higher-dimensional cases can be considered in a similar way. Let $s = (s_1, s_2)$ describe the location of a circuit element in the plane. The model appears to be the same as before:

$$\tau\frac{\partial^2 \theta}{\partial t^2} + \frac{\partial \theta}{\partial t} = S_0 + \omega + \int_{-\infty}^{\infty}\int_{-\infty}^{\infty} K(s - s')f\big[\theta(s, t) - \theta(s', t)\big]\,ds',$$

but now

$$K(s) = \sum_{j=1}^{j=3} \frac{g_j}{2\pi\sigma_j^2}\exp\left(-\frac{|s|^2}{2\sigma_j}\right),$$

where $|s|^2 = s_1^2 + s_2^2$. We suppose that there is a state for this system, say $\theta = \Omega t + \Phi(s)$. The steps proceed as before: First, we linearize about the static state, and then we calculate the Fourier–Laplace transform. The resulting dispersion relation looks the same, but with an important difference: The maximum of p is at k_c, and

$$p(k_c) = \text{max}.$$

The most unstable wave vectors are given by an entire circle: If $k = (k_1, k_2)$, then $k_c^2 = k_1^2 + k_2^2$. Consequently, $p(k)$ is maximum for the entire circle of values

$$k_1^2 + k_2^2 = k_c^2.$$

There are many interesting consequences of this fact.

For example, the term

$$\exp\big[p(k)t\big]\cos k \cdot s$$

is a solution of the linearized problem. The vector k gives the coordinates of the factor normal to the line $k \cdot s = 0$. Thus, this solution appears to be an infinite wavy surface. It has height 1 on the lines $k \cdot s = 0$ (MOD 2π), and its waves are orthogonal to k.

Since an entire circle of wave numbers are most unstable, all of the ingredients needed for circles supporting the unstable pattern are present. Circles having radius $2\pi/k_c$ and packed into the plane are interesting to consider. Figure 7.7 shows that the angle subtended by an adjacent circle's tangent through a radius is 30°. Thus, each circle will have six adjacent neighbors. A plot of these is given in Figure 7.7, and we see that the excluded parts of the plane form a hexagonal pattern! In experiments it would be difficult to tell the difference between real hexagonal structure and closely packed circles. Although all the ingredients seem to be available for making circles, there is not yet a proof that they do indeed represent the modes of growth.

If we carry out the same calculations for $s \in R^3$, we conclude that a spatial pattern of growth will occur on spheres of radius $2\pi/k_c$. However, packing spheres in R^3 is not as easy as for circles in R^2. In fact, there is no unique "best" packing of spheres in R^3 [20].

6.5 Cellular automata

In this section, we consider a two-dimensional array of sites, and rather than considering what happens continuously in time, we will sample this array at discrete time intervals that are somehow natural to the state variables. However, the choice of time step poses a problem for the theory. How should the time step be chosen to be consistent with the physical system? Ignoring this question, we take an arbitrary, but fixed size for the time step.

Consider an array of elements covering a region of the plane and having addresses i, j. Suppose the state of the cell at site i, j is denoted by $S_{i,j}$.

6.5.1 The Game of Life

The Game of Life supplies a good example of this problem. Suppose we have an infinite planar array of sites whose states are either active (1) or inactive (0). At each tick of some clock these states change according to a set of rules suggested by Conway [19]. A site is said to be firing if $S = 1$ and resting if $S = 0$. The rules for change are:

1. A firing site having either two or three firing neighbors will continue to fire. (There are eight neighbors to each site.)
2. Each site having zero, one, or four or more firing neighbors stops firing.
3. Each site starts firing if it is resting and exactly three neighbors are firing.

These rules are chosen to ensure a rich structure of transient and persistent patterns. An entire generation of computer students has devoted untold amounts of computer time to studying this model. Such an approach is not very efficient for despite all this effort the model is still not very well understood. The following diagrams indicate some of the more interesting initial patterns.

a. the blinker (period 2):

$$
\begin{matrix}
0 & 1 & 0 \\
0 & 1 & 0 \\
0 & 1 & 0
\end{matrix}
\rightarrow
\begin{matrix}
0 & 0 & 0 \\
1 & 1 & 1 \\
0 & 0 & 0
\end{matrix}
$$

b. the stable block:

$$
\begin{array}{cc}
1 & 1 \\
1 & 1
\end{array}
$$

c. the stable beehive:

$$
\begin{array}{cccc}
0 & 1 & 0 & 0 \\
1 & 0 & 1 & 0 \\
1 & 0 & 1 & 0 \\
0 & 1 & 0 & 0
\end{array}
$$

d. the glider:

0 1 0 0	0 0 0 0	0 0 0 0	0 0 0 0	0 0 0 0
0 0 1 0	1 0 1 0	0 0 1 0	0 1 0 0	0 0 1 0
1 1 1 0 →	0 1 1 0 →	1 0 1 0 →	0 0 1 1 →	0 0 0 1
0 0 0 0	0 1 0 0	0 1 1 0	0 1 1 0	0 1 1 1

The configuration in a repeats every second step, those in b and c are stable and do not change, and the configuration in d reproduces itself each fourth step, but translated to the lower right; this is an example of a steady progressing wave of activity.

The Game of Life highlights two particular problems. First, the choice of time units is arbitrary. It is not clear how to choose the time step appropriately in a given application. Second, there is no good way to analyze such mathematical systems: Although computer simulation is straightforward, it is quite difficult to predict what will be the eventual fate of a given initial configuration or to begin with an established configuration and to determine what initial configurations could have generated it.

One can visualize this process as taking place on the entire plane, and the evolution of some initial configuration of 1's can be followed. However, it is impossible to use a computer in an unbounded domain unless some restrictions are made. Therefore, we consider a finite region of the plane.

The question arises of what happens at the boundaries of the domain; namely, what boundary conditions are appropriate for the model? Periodic boundary conditions are the easiest to set. In this, the problem is assumed to be periodic in both directions, say having period N. Then $S_{i+N,j} = S_{i,j} = S_{i,j+N}$, etc. for all i, j. With these boundary conditions, it is only necessary to consider a fundamental domain, $0 \le i, j \le N$. An interesting rendition of this problem results: Cut out the fundamental domain, identify the top and bottom by sewing them together, and then identify the right and left edges by sewing these together. The result is a torus! The grid points (i, j) now label points on the surface of a torus.

If you consider the glider with these boundary conditions, it will successively pass through the fundamental domain, moving out one side and in through the opposite one.

The array can be visualized in several other ways using other boundary conditions. For example, it might be a finite array that has *reflecting boundaries*; that is, there is a patch of side N such that the entire network is described by the addresses $i, j = 0, \ldots, N$, and false position sites are introduced at $N + 1$ with the state at that site being equal to the state at the same site but with the index $N + 1$ replaced by $N - 1$. Alternatively, the boundaries might be *absorbing*, and we simply always set $S_{i,j} = 0$ if (i, j) lies on a boundary.

6.5.2 Networks having refractory elements

The topics described in this section will not play a direct role in our brain models, but they are interesting for historical reasons and have been considered as models for networks of neurons. In addition, they might suggest some things for us to pursue in general network models.

The Game of Life is fun to play with, and it provides a good introduction to the idea of discrete event modeling of spatial spread of activity. But it is quite distant from real applications to neural networks. Each site should have at least three possible states, say excited, resting, and refractory, and interaction with a single neighbor should be allowed. Therefore, we now suppose that each site can have three possible states: $S = 1$ (excited), $S = 0$ (resting), and $S = -1$ (refractory), and we can construct rules for changes of state. The states change at each tick of a clock according to the following rules:

1. An active site becomes refractory $(1 \rightarrow -1)$.
2. A refractory site becomes inactive $(-1 \rightarrow 0)$.
3. An inactive site becomes active if any of its four nearest neighbors is active.

Of course, these rules are no easier to work with than the Game of Life, but the kinds of behavior possible in this system are very interesting. Models of this kind have been used to model cardiac dynamics where each site corresponds to a muscle cell (see [56]). The mechanisms for neural networks are mathematically similar to those of cardiac dynamics, since both are based on the Hodgkin–Huxley model. This model is referred to as the Wiener–Rosenblueth model.

A single excited site will create a wave of excitation that propagates out from the initial site, as shown in Figure 6.3. The wave of activation is followed by a refractory wave. If this array were on a sphere with the obvious identification of neighbors at the poles, then the wave of excitation would collapse antipodal to where the stimulation was applied.

$$
+ \rightarrow
\begin{array}{c}
+ \\
+ - + \\
+
\end{array}
\rightarrow
\begin{array}{c}
+ \\
+ - + \\
+ - 0 - + \\
+ - + \\
+
\end{array}
$$

Figure 6.3. Propagation of excitation in a cellular automaton.

There are two particularly interesting initial configurations that lead to repeating patterns:

1.
$$
\begin{array}{c}
+ - 0 \\
0 - +
\end{array}
$$

This initial pattern generates a repeating pattern of period 3. It is the highest frequency pattern generator known and was discovered by Platt [110].

2.
$$
\begin{array}{c}
+ - \\
- +
\end{array}
$$

This pattern generates a repeating pattern of period 4 and has been studied by Greenberg and Hastings [41].

Each of these initial patterns repeatedly generates waves that propagate outward from the initial site and have wavelength equal to their period. When two series of wave patterns intersect, the shorter one (the higher frequency one) eventually takes over the entire array. Spiral waves similar to these are observed in certain chemical reactions, such as the Belousov–Zhabotinsky reaction [142], and they are implicated in cardiac fibrillation.

6.5.3 Ising–Hopfield model

J. J. Hopfield [53] proposed a model for studying large networks of nerves in a way that is reminiscent of the Ising model from theoretical physics [101].

We consider a collection of N circuit elements (where N is large) that are either on ($+1$) or off (0). Let v_j denote the state of the jth element, and let $\mathbf{v}$ denote the vector of these states. Suppose that the number $T_{i,j}$ describes the strength of connection from site i to site j. (Note that we take $T_{i,i} = 0$.) Let T denote the matrix of these connections. Then the product $T\mathbf{v}$ describes the net input within the network; for example, the ith component of this product is $(T\mathbf{v})_i = \sum_{j=1}^{N} T_{i,j} v_j$, which describes the net input to site i from the system. The states change dynamically according to rules, as in the Game of Life, but now the sampling of inputs decides what changes are made and these occur at random time intervals. This randomness results in the system becoming asynchronous in time, making the graphical methods in the preceding section not directly applicable.

The algorithm. Suppose that we are to decide what happens at site i; that is, the time has come to sample element i's input to update its activity. We specify the following rules:

1. If $(T\mathbf{v})_i > 0$, then $v_i \to 1$.
2. If $(T\mathbf{v})_i = 0$, then v_i does not change.
3. If $(T\mathbf{v})_i < 0$, then $v_i \to 0$.

These rules describe the changes that will occur when sampling each site. The next question is: When is the ith element sampled? The following rule establishes the sampling order:

Draw N numbers, say $\{\tau_1, \ldots, \tau_N\}$, according to some probability distribution, say having mean value $1/w$. w is called the mean sampling rate.

The smallest of these sampling times tells which component (say i) is sampled first; that is, $\tau_i \leq \tau_j$ for $j \neq i$. Apply rules 1, 2, and 3 to site i, draw another random number from the same distribution, say τ, and replace τ_i by $\tau_i + \tau$ in the list of sampling times. Now, find the next sample time in the list and repeat these steps.

This asynchronous sampling scheme circumvents the problem of using an arbitrary fixed sampling interval as in the Rosenblueth–Wiener model.

An energy function. An energy surface of some sort would be useful in describing this complicated model. Hopfield (see [53]) defines an energy function by the formula

$$E(\mathbf{v}) = -\frac{1}{2}\mathbf{v}^{tr}T\mathbf{v},$$

where $\mathbf{v}^{tr}$ denotes the transpose of the vector $\mathbf{v}$. Minima of E will correspond to stable firing patterns for the network much like the minima of F describe the stable equilibria for x in the preceding section. Unfortunately, the surface $E = E(\mathbf{v})$ cannot be drawn since there are N independent variables $v_1, \ldots, v_N$, where N is large.

Hopfield's model and short-term memory. Suppose that we want some state $\mathbf{v}^*$ to be a stable configuration for the algorithm. That is, $\mathbf{v}^*$ should be reproduced by the algorithm and several other states should eventually reach $\mathbf{v}^*$ under the algorithm. How can the connection matrix T be chosen to ensure that this is

Table 6.1. *Dynamics of Hopfield's model for* $N = 4$

Energy				
2		1111		
1	1011	1010	0111	
	0011	0101	1110	
	1101	1100		
0	1000		0100	
	0001		0010	0000
−1	1001		0110	
			stored	

the case? Hopfield suggests that

$$T_{i,j} = (2v_i^* - 1)(2v_j^* - 1)$$

works in some sort of average sense.

Let us investigate this model in a particular case, say where $N = 4$. In this case, there are 16 possible states: 0000, 0001, ..., 1111. Suppose that we want to store 0110. To do this, we set

$$T = \begin{pmatrix} 0 & -1 & -1 & 1 \\ -1 & 0 & 1 & -1 \\ -1 & 1 & 0 & -1 \\ 1 & -1 & -1 & 0 \end{pmatrix}.$$

Table 6.1 summarizes the results in this case. This calculation shows that there are three static states: the stored state 0110, its complement 1001, and the dead case 0000. The column headed by 1111 consists of states that can evolve into either 0110 or 1001 depending on the outcome of sampling. Nothing evolves into the dead state. Note that the algorithm preserves not only the desired "word," but also its complement obtained by replacing 1 by 0 and 0 by 1 everywhere.

One feature of this model is that it shows how to connect sites to store a given word. A drawback of it is that analysis of the general case is very difficult.

A great deal of work, especially over the past 15 years, has been devoted to the development and use of artificial neural networks. These usually are composed of several layers of circuit elements, often an input layer, a hidden layer, and an output layer, with a general network of connections initially emplaced between contiguous layers. The network is trained in the following way: First, a vector of inputs, say **I**, is applied to the input layer and we intend to get a vector of outputs, say $\mathbf{O}_d$, for the desired output. The actual output is observed, say **O**, and the connections are then modified to get the actual output as close as possible, for example, in the mean square sense, to the desired output. There are various algorithms for modifying the connections in a systematic way, including back-propagation and similar gradient descent methods [120].

The network is trained by repeating this process for a set of training vectors $\{\mathbf{I}_j, \mathbf{O}_j\}$, and then it is turned loose to deal with arbitrary inputs; often the results are quite good for pattern recognition.

Since this line of work is not in the direction of our analysis of neural networks, we do not pursue it further here.

6.5.4 Markovian models

Another important way to describe a network is based on the probabilities of transitions from one state to another. Consider a general system characterized by a state S that can occur in M possible ways, say $\{S_1, \ldots, S_M\}$. We sample the state of the system stroboscopically at each tick of some clock, and so observe a sequence of states

$$S_{i_1} \to S_{i_2} \to S_{i_3} \to S_{i_4} \to \cdots,$$

where $i_1, i_2, \ldots \in \overline{1, M}$.

From these observations, we can build up a table of probabilities of transitions, say

$$P_{i,j} = Pr[S_i \to S_j],$$

which we summarize as a matrix P. Then we can describe the system not by its present state, but rather using the probability distribution that we should observe over many experimental observations: Let $\mathbf{p}_n = (p_{1,n}, \ldots, p_{M,n})$, where

$$p_{i,n} = Pr[S = S_i \text{ at the } n\text{th clock tick}].$$

The vector $\mathbf{p}_n$ is the probability distribution of states at the nth sampling. The sequence of vectors $\mathbf{p}_0, \mathbf{p}_1, \mathbf{p}_2, \ldots$ then describes the outcomes of what we should see in many repeated trials (experiments). It is possible to derive a model for the probability distribution vectors:

$$\mathbf{p}_{n+1} = \mathbf{p}_n P,$$

where we can think of the vector $\mathbf{p}_n$ as being a row vector and P as the matrix of transition probabilities. (The reason row vectors are used is to preserve the mnemonic that $P_{i,j}$ is the probability of moving from state i to state j in one time step.)

The random process described here is a *Markov chain*, and a great deal is known about such processes [25]. First of all, the states can be divided into two

types: transient and recurring. Recurring states are those that once visited are
returned to again in the future, and transient ones are the others. In fact, the
matrix P can be rewritten, by relabeling the states if necessary, to the form

$$\begin{pmatrix} R & 0 \\ T_R & T \end{pmatrix},$$

where R describes transitions among the recurring states and T describes the
probabilities of staying in the transient states, which are all less than 1. This is
reflected in the fact that all of the eigenvalues of T satisfy $|\lambda| < 1$. Note that
row sums of this matrix must equal 1 since the probability of moving to or
remaining in *some* state is one.

In addition, the structure of R is interesting and useful. It can be written as
a diagonal matrix of matrices:

$$R = \begin{pmatrix} R_1 & 0 & 0 & 0 \\ 0 & R_2 & 0 & 0 \\ 0 & 0 & \ddots & 0 \\ 0 & 0 & 0 & R_k \end{pmatrix},$$

where one eigenvalue of each of the matrices R_j is 1 and the corresponding
eigenvector is full in the sense that all of its entries are nonzero. Thus, each
state is accessible to the process, and the components of this matrix describe
the probabilities of eventually being in the respective state. The matrices R_j are
called *irreducible*, and they describe the recurrent states of the process. Each
one describes a collection of states that communicate with one another; once
in such a collection, the state never leaves. Each collection of states for $\mathcal{R}_j$ is
referred to as being *ergodic*.

One might make the analogy that each ergodic collection describes a program
of activity that is memorized by the process, and once the network is initialized,
say with an initial vector

$$\mathbf{p}_0 = (\delta_{1,I}, \delta_{2,I}, \ldots, \delta_{N,I})$$

where $S_I \in \mathcal{R}_I$ is the initial state of the process, then the process continues
in this collection of states and executes the corresponding sequence of steps
randomly as prescribed by the transition matrix.

Markov chain model in a random environment. Suppose now that the Markov
chain just described is perturbed by a small random fluctuation due to noise in
the system's environment. Then the transition matrix is replaced by a sequence
of matrices

$$P_\varepsilon(n) \equiv P + \varepsilon Q(n)$$

for $n = 0, 1, 2, \ldots$, where the row sums of $Q(n)$ are zero and the matrix P_ε has nonnegative entries whose row sums are 1. This sequence can on average be written in terms of components in the form

$$
P_\varepsilon(n) = \begin{pmatrix}
R_1 & \varepsilon Q_{1,2}(n) & \cdots & \varepsilon Q_{1,k}(n) & \varepsilon Q_{1,T}(n) \\
\varepsilon Q_{2,1}(n) & R_2 & \cdots & \varepsilon Q_{2,2}(n) & \varepsilon Q_{2,T}(n) \\
\varepsilon Q_{i,2}(n) & \cdots & \ddots & \cdots & \varepsilon Q_{i,T}(n) \\
\varepsilon Q_{k,1}(n) & \varepsilon Q_{k,2}(n) & \cdots & R_k & \varepsilon Q_{k,T}(n) \\
\varepsilon Q_{T,1}(n) & \varepsilon Q_{T,2}(n) & \cdots & \varepsilon Q_{T,k}(n) & \varepsilon Q_{T,T}(n)
\end{pmatrix}.
$$

As a result of this averaging, we see that the probability of leaving an ergodic collection of states is small (order ε) and that the expected residence time in the collection is of order $1/\varepsilon$. Such processes are studied in [122].

Chetaev has studied models of this kind in neuroscience. In particular, he has studied bifurcation behavior in such networks [15].

6.5.5 Large deviation theory

The work of Ventzel and Friedlin [136] has interesting implications for our discussion of dynamics of gradient systems. They considered such problems when the system is perturbed by a Wiener process: Consider the model

$$
dx = -\nabla F(x)dt + \sigma dW,
$$

where $x \in R^1$ and dW is white noise. This equation can be analyzed in R^1 like the Markov chain was in the preceding section. The analogy of $\mathbf{p}_n$ in this case is a probability distribution function $\phi(x, t)$ that describes the probability of the solution being near the point x at time t. This function is known to satisfy the Kolmogorov equation

$$
\frac{\partial \phi}{\partial t} = \frac{\partial}{\partial x}\left(\frac{\sigma^2}{2}\frac{\partial \phi}{\partial x} + \phi \nabla F\right).
$$

The equilibrium distribution is obtained by solving

$$
\frac{\sigma^2}{2}\frac{\partial \phi}{\partial x} = -\phi \nabla F,
$$

and the solution is

$$
\phi(x) = \exp\left(\frac{-2F(x)}{\sigma^2}\right).
$$

If σ is small, then the probability of being near x will be largest at values of x where $F(x)$ is smallest. Again, the minima of F appear as the most likely places for the solution to be residing. However, it is not certain that solutions will remain near a minimum of F. In fact, they probably leave each potential well they enter; however, they are expected to reside in a well for a length of time proportional to

$$\exp(d/\sigma),$$

where d is the depth of the well, and the expected time it spends between wells is of order σ. Thus, if the variance of the noise (σ^2) is sufficiently small, then the system will usually be in an energy well for a long time, and it will migrate between wells for a short time.

Wentzel and Friedlin also obtained a Markov chain approximation to the system to describe transitions that solutions will make between wells.

6.6 Summary

This chapter presented some methods useful for studying large networks. Of particular importance are systems that are either gradient systems or are like gradient systems in the sense that there is a Liapunov function for them. The underlying potential or Liapunov function is referred to in this context of brain modeling as defining a mnemonic surface.

Mnemonic surfaces can be formulated, uncovered, and embedded in networks in a variety of interesting and useful ways. But the actual mechanisms of memory are still not understood [76].

Bifurcations are important phenomena for us to pursue in studying networks. These are often accessible through experiments, and there is a well-developed mathematical methodology for studying them. (See [74] for further work in this direction.)

We showed how general integral equation representations of networks can help with stability analysis. In particular, it was shown how patterns of firing can be described in terms of most-unstable wave numbers. We will use this analysis and this kind of model in the next chapter.

Cellular automata can be formulated and used to study networks. They are easy to simulate, although difficult to analyze. They do point to various things we can look for in more realistic networks, and we will use the results obtained in this chapter later. There are continuous versions of some cellular automata that have been successfully studied (e.g., see [85]).

Finally, we have seen that there are many ways that random noise can be folded into our models, and there are well-developed mathematical methodologies for analyzing those cases. Particularly important are the Law of Large Numbers and the Central Limit Theorem, which underlie most stochastic methods. We have seen how noise can play a role in dynamics superimposed on mnemonic surfaces.

6.7 Exercises

1. *Game of Life on a torus.* Computation with the Game of Life is not possible in a computer unless some restrictions are placed on the domain. Describe the dynamics of the Game of Life when the x-y plane is taken to be periodic in both directions. Specifically, suppose that the state at the site (i, j) is identical with the state at sites $(i + N, j)$, $(i, j + N)$, and $(i + N, j + N)$. Show that this is equivalent to considering the Game of Life on a grid drawn on the surface of a torus. Compute the evolution of the glider in this case.

2. *A single propagating wave in a refractory system.* Show that a single excited site generates a propagating wave of excitation followed by a refractory wave by iterating the model in Section 6.5.2 geometrically.

3. *Spiral waves in refractory networks.* Show that the spiral waves in Section 6.5 continue by carrying out an appropriate number of iterations until the central region repeats itself.

4. *Networks having refractory elements in three dimensions.* The model described in Section 6.5.2 can be studied in three dimensions. Let $S_{i,j,k}$ denote the state 1, 0, or -1 at site (i, j, k). The rules for changing states are as before: (a) $+1 \rightarrow -1$; (b) $-1 \rightarrow 0$; and (c) $0 \rightarrow +1$ if any of its six nearest neighbors is $+1$. Define a grid of $20 \times 20 \times 20$ sites, and impose periodic boundary conditions in all three directions. Simulate this model using the initial configuration $S_{10,10,10} = +1$, $S_{10,10,9} = -1$. Describe the evolution of this initial configuration.

5. *The wall.* Determine the behavior of the initial configuration
   $$\cdots 000000 + + + + + + + + + + + + + \cdots$$
   $$\cdots 000000 - - - - - - - - - - - - - \cdots$$
   in the system of Section 6.5.2. Here $\cdots$ indicates that the symbol is repeated forever.

6. *Gradient system.* Show that if $\mathbf{x}(t)$ solves the differential equation
   $$\mathbf{x} = -\operatorname{grad} F(\mathbf{x})$$
   and starts near a minimum of F, then $\mathbf{x}(t)$ approaches a minimum of F. In particular, show that there can be no periodic solution of this equation by showing that the integral of $(d\mathbf{x}/dt)^2$ around an orbit is zero.

7. *Stored memory in Hopfield's model.* Carry out the computation in Section 6.5.3 when the stored state is to be 1101. You should find that 1101

(energy $= -3$) and 0010 (energy $= 0$) are stable static states. Also, show that the states 1010 and 0110 can evolve into either stored state.

8. *Asynchronous Game of Life.* Use Hopfield's asynchronous sampling scheme in the Game of Life. Using a computer simulation, determine what happens to the configurations shown in the diagrams in Section 6.5.1.

9. *Gaussian kernel.* Consider the kernel in Section 6.4.3. Show that the data $g_1, g_2, g_3, \sigma_1, \sigma_2, \sigma_3$ can be chosen so that $K(s)$ is positive for all s, but $\tilde{K}(k)$, its Fourier transform, has a negative value at its minimum.

10. *Dispersion relation of a Gaussian kernel.* Reproduce Figure 6.1.

11. *Packing spheres.* How many nearest neighbors does a sphere have in a three-dimensional packing of them?

7

Attention and other brain phenomena

How does the brain work? It is known that the brain has a very rich facility for reliably storing, processing, and recalling data, and it can sustain a variety of complicated activities, apparently spontaneously and simultaneously. Many mechanisms for doing these things are known, but most of the ways in which the brain performs these processes are unknown.

Important investigations of brain response have been carried out by generations of physiologists and psychologists. Two particular aspects of this work have been the collection of vast amounts of data about how the brain responds to various inputs and the formulation of models for how it responds to stress and stimulation. Helmholtz, Freud, Jung, and others found it useful to associate with brain activity an energy surface whose valleys correspond to (stable) repeatable brain responses. In such a model, stimulation from outside the brain's neural network initializes it, and so locates a starting point on the surface. The brain's firing pattern will then change dynamically, and it will evolve into a stable pattern of frequencies and phase deviations that corresponds to a valley floor on the surface.

A stable firing pattern is the brain's response to the stimulus; the organism reads out this firing pattern, for example, by modulating respiration or causing release of hormones or endorphins. This stable pattern may short lived, soon replaced by other patterns as we saw in the sound localization process in Chapter 5.

We earlier used the rotation vector method to show that network models in the frequency domain have solutions of the form

$$\theta = \Omega t + \phi(t),$$

where the phase deviation ϕ is determined by solving a gradient-like system. As before, we view the frequency as identifying a channel (or station) for commu-

153

nication and ϕ as carrying information contained in the potential function. The potential function is shaped by the network and the relationship and the network of connections between network elements. This important topic is discussed in greater detail than presented here in [67].

In this chapter, we turn to descriptions and simulations of some known brain networks that have been uncovered through experiments and that carry out specific tasks in the body. These involve focusing attention, propagating activity in the neocortex, and development of some properties of the visual cortex. Finally, we demonstrate that such networks can support waves of activity propagating through them.

This book is an introduction to mathematical methods for the neurosciences, and that continues to be our focus. However, it is at this point that we can start applying the work in earlier chapters to problems in brain science, and this chapter is intended to be a launching point. We derive some basic models, briefly discuss each, identify some things of interest about them, and suggest directions for future study.

We first study the *searchlight hypothesis* for modeling how a network can focus attention. This network allows to pass only the most persistent (highest frequency) signals, and it is motivated by observations of how the thalamus and reticular complex interact. We use computer simulation to illustrate this circuit's behavior.

Next, the structure of a cortical column is described. Our model describes a collection of excitatory neurons in the column's medulla and a source of inhibition deep down in the structure. We study the frequency and phase response of a column by computer simulation. The behavior of a column is quite similar to that of the atoll model, which is the basis of our model for attention, in that there is reciprocal interaction between excitatory and inhibitory elements. The new feature here is that the oscillators can be modulated by outside influences in more complex ways. Interacting excitatory and inhibitory groups of neurons are referred to as forming a *neural oscillator* [67, 140]. We return to this structure in our study of wave propagation at the end of this chapter.

Columns are connected to form a model of a region of the neocortex that is essentially a laminated sheet seven layers deep; the first six layers are excitatory and the seventh (deepest) is inhibitory. Computer simulations show that excitation of one column can recruit or inhibit activity in other columns. The model is simplified successively by lumping column parts into units. First, the full column model is considered. Then a double-layer model, with one layer excitatory and the other inhibitory, is described. Finally, a single layer of VCONs that describes oscillations between the excitatory and inhibitory elements is developed. These models lay the basis for a number of simulation projects.

Next, we describe learning in a large network by modeling the development of the visual cortex in newborns. There results from stimulation through the eyes a pattern of connections between elements in the model visual cortex describing which columns are dominated by inputs from which eye. This pattern is developed through strengthening of connections through correlated activity, and it is (quasi-) permanently stored in the connections.

Our notation for three-dimensional arrays of circuits is described by an addressing scheme that associates with each location a triplet of integers, say (i, j, k). There are 26 neighbors to an address in a rectilinear array and 6 nearest neighbors; various coupling stencils for neural networks are described later in this chapter. However, it is often convenient to use other listings of these addresses, such as a single index that gives a serial, or lexicographic, listing of the circuit elements. In other settings, we will use other lexicographic conventions. In studying large networks it proves useful to convert the addressing into a system using a continuous variable, say $s = (s_1, s_2, s_3) \in R^3$. This enables us to use methods of calculus to uncover wave propagation phenomena in large networks, which we do in the last section of this chapter.

7.1 Attention: the searchlight hypothesis

The cocktail party problem illustrates how we focus attention on one of several competing stimuli: During a conversation with one person, your attention might be diverted if someone near you, outside your group, mentions your name. In that case, your attention is diverted to the new stimulus. How does this happen?

It is believed that stimuli get to the brain's neocortex, where further processing is done, through the thalamus. From the neocortex, they are passed to the hippocampus for further processing and memorization and then sent back to lower brain centers to modulate activity, respond, etc. [21, 95]. It has been proposed that the thalamus acts like a searchlight, picking out the most important signals and allowing them to pass while blocking others. When a stronger signal arrives, the circuit shuts down the current focus of attention and recognizes the thalamic output to the new stimulus. Stimulus intensity is programmed in the mind by experience and genetics. Having a hierarchy of attentional priorities enables us to meet threats, to take opportunities for food, to deal with injuries, etc.

We derive here a VCON network that does this, which is based on a discussion in [21, 59], and we are able to produce a searchlight-like tracking mechanism. In rough terms, a signal is received by a thalamus cell, and it excites a cell in the reticular complex. The thalamus is made up of excitatory cells that are not coupled, and the reticular complex (RC) is a network of inhibitory cells that are coupled laterally. The RC cells also feed back to the stimulating thalamus cell and feed forward to specific columnar regions in the neocortex

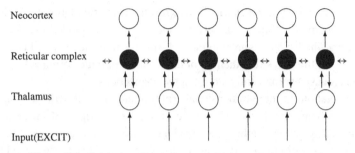

Figure 7.1. Depiction of a thalamus-reticular complex network.

where they inhibit an inhibitory cell, and so are functionally excitatory in the neocortex.

Crick [21] cites evidence that inhibition from the reticular complex cell causes the thalamic cell to fire a burst of action potentials, rather than to turn it off. The idea is that when a dominant stimulus arrives, a thalamic cell is stimulated to fire a burst; the corresponding RC cell shuts off the neighboring cells, closing channels for their thalamic cells, and at the same time electrical activity is created in the corresponding column of cortical cells.

In the model presented here, the stimulus (frequency σ) excites a thalamic cell, which in turn excites an RC cell. The cell in the RC inhibits neighboring RC cells and feeds back to inhibit the driving thalamic cell. The inhibition of the thalamic cell, rather than stopping it from firing, causes it to fire a burst for a fixed time, and then become quiet, much as in the atoll model. The atoll oscillator provides a basis for modeling the thalamus and RC cells since slow inhibition from one causes rapid bursting of the other.

We suppose that the slow reticular complex cell inhibits its neighboring RC cells, its thalamic cell, and inhibitory cells in its cortical column. Thus, the reticular complex cell, while being an inhibitory cell, is functionally excitatory in the neocortex. This is sketched in Figure 7.1.

The following VCON network is functionally equivalent to the one depicted in Figure 7.1. We model the cortical column as being an excitable VCON and a Grand Inhibitor (GI) VCON. In our model, the RC cell feeds inhibition to the GI. Thus, when one of the RC cells fires, the corresponding cortical column fires a burst as well. This activity is our measure of output of the cortex. We take purely inhibitory connections among all cells in the reticular complex: We suppose that each cell inhibits all neighboring cells in either direction along the RC. Most other inhibitory stencils should work as well. Finally, we ignore feedback from the neocortex and other systems to the RC and to the thalamus.

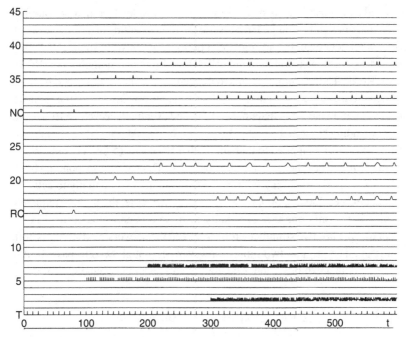

Figure 7.2. Simulation of the attention model as described in text.

Consider an infinite array of these circuits into which identical inputs are placed every N cells. Thus, we need only consider N cells in a periodic array, and the mathematical equations for the model are

$$\dot{x}_j = 1 + \cos x_j + 0.95 \cos_+ x_{j+N}, \tag{7.1}$$

$$\dot{x}_j = 0.04 \left[1 + \cos x_j + S\left(2 \cos_+ x_{j-N} - \sum_{k=N, k \neq j}^{2N-1} \cos_+ x_k \right) \right], \tag{7.2}$$

$$\dot{x}_j = 5(1.0 + \cos_+ x_j - \cos_+ x_{j+N} + \sigma_j). \tag{7.3}$$

In the first equation, $j \in \overline{2N, 3N - 1}$ in the second $j \in \overline{N, 2N - 1}$, and $j \in \overline{0, N}$ in the third.

The connection matrix in the summation in the function $S(u)$ describes inhibitory connections from the other RC VCONs impinging on the jth cell. The sum in the RC equation is taken over the RC layer ($k \neq j$ and $k = N$ to $k = 2N - 1$).

Figure 7.2 summarizes the following computer experiment. We consider the network with $N = 15$ cells in each of 3 layers: the thalamic VCON (T0, ...), the RC VCONs (RC15, ...), and the neocortical VCON (NC30, ...). These

are wired as described earlier. In particular, we consider 15 reticular complex
and 15 thalamus cells arranged in a rosette [57], representing an infinite array of
cells that are identical, but arranged in a periodic array. We visualize thalamus
cell T14 as being next to thalamus cell T0, and reticular complex cell RC15
next to RC29. The incoming signals to the thalamic layer are described by their
forcing activities σ_i. First, activity 0.01 is applied to VCON T0. To the right we
see oscillatory behavior in thalamic cell T0, and eventually a cycle in RC15 and
a corresponding burst in the cortical layer cell NC30. At time 100 a stimulus of
strength $\sigma_4 = 0.1$ is applied to thalamus cell T5. T0 and T5 fire, but only RC20
eventually fires; the other RC cells are quenched. There is a corresponding burst
in NC35. Activities of strength 5.0 and 6.0 are applied sequentially to cells T7
and T2, at times 200 and 300, respectively. In each case, the higher-frequency
stimulus is the only one that eventually leads to a burst in the cortical layer,
except that the last two stimuli are too close, and the network reads them as both
significant. The original two inputs, however, are not allowed to pass. In this
sense, the model exhibits winner-take-all dynamics in the frequency domain.

7.2 Column structures in the neocortex

The neocortex is made up of column structures that are aligned orthogonal to
the surface, and they are striated into six layers. We model these structures in
this section using VCONs to describe network elements. This model begins
with a formulation of a network arranged like the neurons observed in cortical
columns. We make the following assumptions:

- The regions we model are arranged in columns orthogonal to the surface of
  the neocortex and project into the cortex.
- Columns comprise mostly excitatory cells and inhibitory interneurons.
- The excitatory cells are striated on six layers (I–VI), although we ignore
  Layer I [81].
- There are projections from the excitatory cells deeper into the cortex where
  we suppose that they connect to an inhibitory cell, called the Grand Inhibitor
  for the column.
- The Grand Inhibitor projects back up to inhibit the excitatory neurons in the
  medulla.
- Excitatory cells have only excitatory synapses, and inhibitory cells have
  only inhibitory connections. This is sometimes referred to as being Dale's
  principle [74].
- Inputs to the column can be either excitatory or inhibitory, and may arrive
  either to the Grand Inhibitor or to all cells on Layer IV.

- The Grand Inhibitor can project synapses to other columns, both to the other's Layer IV and to its Grand Inhibitor.
- Layer IV can project synapses to other columns, both to the other's Layer IV and to the Grand Inhibitor.
- Inputs might arrive from deeper in the brain, for example, from the reticular complex.
- Inputs might arrive from outside the cortex to Layer IV.
- Outputs might be to other brain columns or to other brain structures such as the hippocampus, the motor cortex, etc.

For the most part, we model the arrangement of cells on each layer of a column to be in a circularly symmetric arrangement that we refer to as being a *rosette*.

7.2.1 Cortical columns

The visual cortex of most mammals is populated by columnar structures of neurons aligned orthogonal to the surface of the neocortex. These columns are (in rough terms) 3-mm long and contain about 200,000 neurons. These are usually striated into six layers [69, 81]. A cortical column, sometimes referred to as being a barrel structure, comprises a medulla having six layers with mostly unknown connections between a variety of known and unknown neurons [70]. The first layer, sometimes referred to as being the tangential layer, contains mostly dendritic material, and we do not account for it here. We do model the other five layers (II–VI) as being occupied by excitatory cells that communicate with a deep inhibitory cell, called here the *Grand Inhibitor* (or GI) [9, 10].

One can visualize a handful of wooden pencils with the erasers being the top layer of the neocortex. If one paints on each pencil five bands of different colors (blue, red, . . .), the result would be a handful of pencils showing six layers – the eraser layer, the blue layer, the red layer, etc. A column has six layers on which reside a configuration of excitatory cells. They have many interconnections and they have processes that go deep into the cortex. We assume that this deep penetration is to an inhibitory cell, residing at the pencil tip, that feeds back to the rest of the column. We refer to this as the *pencil model* of a column.

It is not known why there are many neurons in a column that seem to respond in similar ways to stimuli. It has been speculated that redundancy in the network makes it more efficient to reliably process information where the response of a single neuron would be more susceptible to noise in the signal, in connections, and in its own metabolism.

How many columns there are and what they do are not known either. It is speculated that each column responds to single (possibly unique) stimulus.

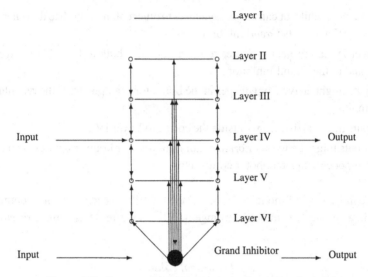

Figure 7.3. Cross section of the cortical column pencil model.

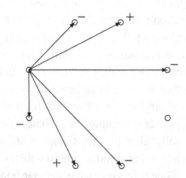

Figure 7.4. Connection stencil for a rosette.

Some fire when the eye sees a straight line in a particular orientation, some fire when a limb is to make a specific move, some fire when a particular facial whisker is stroked, etc.

7.2.2 The pencil model of a cortical column

Figures 7.3 and 7.4 depict a model that captures some ideas about columns in the visual cortex. A column has six layers on which reside a configuration of excitatory cells (o). They have many interconnections and they have processes that go deep to an inhibitory cell (•) that feeds back to the rest of the column.

In Figure 7.3 there is one inhibitory cell (• called the Grand Inhibitor), and each of the six layers is depicted in cross section. The Grand Inhibitor inhibits all other cells in the pencil, and each of the others excites the GI. In addition, each cell excites the cells in the same position on the layers above and below. We ignore Layer I in this model.

Figure 7.4 depicts the connection stencil for a rosette of neurons on a layer. In this case, eight excitatory cells are depicted (○) and the connections from one to the others are shown. Some of the connections are negative, indicating that the connection is functionally inhibitory even though the driving cell is excitatory. This can occur when an inhibitory inter neuron lies between the sending and the receiving cells. Each neuron in the array has this connection stencil to others as well, so the entire rosette network can be visualized by rotating and printing this stencil eight times. In the simulations below, the number of cells on each layer is usually taken to be six.

Each layer is connected with its neighboring layers by vertical excitatory projections to the corresponding cell on the receiving layer. Columns receive stimulation to Layer IV where inputs are distributed identically to all cells on the layer and to the GI. The GI cell receives excitation from all cells in the medulla, and it feeds back to inhibit all of the cells in the medulla.

This model can be visualized using a simple lead pencil as suggested before, where the point is the GI and the lead in the center carries excitatory processes down to the point and inhibitory processes up away from the pencil point to the medulla.

Each cell can be modeled by a VCON, and each layer modeled as being a rosette of six VCONs, that is, a circular array of VCONs connected by a permutation matrix of connection strengths within the layer.

In the following discussion and in the exercises we consider the response of a single column to various inputs; then we consider several possible interactions between two columns. Finally, we consider two-dimensional arrays of columns and their responses to various forcing. A more complete investigation of this is given in [75].

7.2.3 VCON simulations of one column

We consider here a pencil model having 5 layers of rosettes, each having 6 VCONs, and a Grand Inhibitor. Thus, there are 31 VCONs in the model.

Each VCON in a column model is described by a single phase variable, say θ_i for $i = 0, \ldots, 30$, and the voltage output of each is $V(\theta_i)$, where we

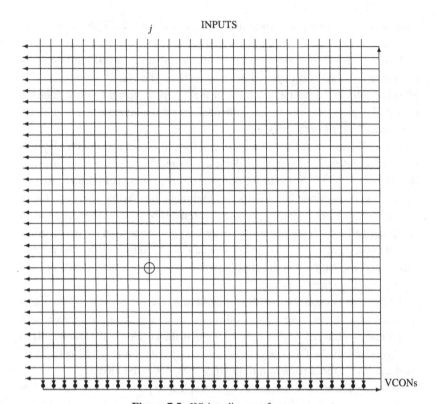

Figure 7.5. Wiring diagram format.

take $V(\theta) = \cos_+ \theta$. Connections within a column are described by a single matrix C having dimensions 31×31. The component C_{ij} gives the strength of connection from VCON i in a column to VCON j for $i, j = 0, \ldots, 30$.

Figure 7.5 shows a format for a wiring diagram for this model; the circle indicates an excitatory connection from site i to site j. Crossing lines are not connected unless there is an open circle (o) or solid dot (●). This diagram shows a network having only one connection, and that is from site i to site j.

The circuit outlined in Figure 7.6 shows a row of 31 VCONs at its bottom represented by small dots. Each VCON has an input coming to it from above and an output emanating from it below. The inputs come from the vertical lines to each. Inputs from outside the network will be indicated at the top of the diagram and inputs from other network VCONs will be indicated by symbols at intersections of horizontal lines with the input line. The VCON outputs combine in a cable that goes from left to right, and at its end, it turns up. The output lines are peeled off the cable and are depicted as horizontal lines going to the left from the vertical cable. The order of outputs is determined by reversing the

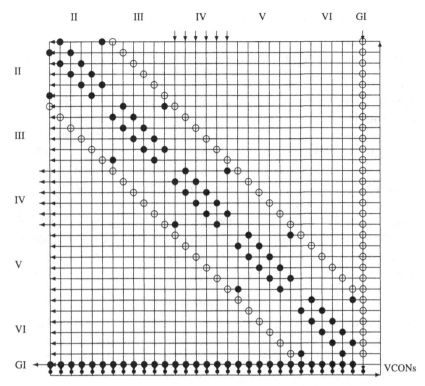

Figure 7.6. The wiring diagram for a single-pencil model of a cortical column. An open circle (o) denotes an excitatory connection, a solid dot (•) denotes an inhibitory connection, and a bare intersection denotes no connection (that is, if there is neither a circle nor a dot, then there is no connection).

order in which VCONs joined the cable: The bottom horizontal line is output from the last (right-most) VCON and the top horizontal line is output from the first (left-most) VCON. The arrows indicate inputs and outputs. The ith VCON puts out a line that reemerges as a horizontal line in position i from the top. The VCONs are labeled $0, \ldots, 30$ from left to right. The Grand Inhibitor is the right-most (number 30), the first 6 are from Layer II, etc. A connection from VCON i to VCON j is indicated on the wiring diagram as being a circle or a dot. Unless a circle or a dot appears at a wiring crossing, there is no connection. An open dot (o) indicates an excitatory connection, and a closed one (•) indicates an inhibitory connection.

The VCON column model is

$$\dot{\theta}_i = 1.0 + V(\theta_i) + \sum_{k=0}^{30} C_{k,i} V_+(\theta_k) \qquad (7.4)$$

for $i = 0, \ldots, 30$.

Our labeling convention will be:

$i = 30$ is the Grand Inhibitor.

For the other $i \in \overline{0, 29}$, their layer number is Int$[i/6] +$ II. Thus, VCONs 0, 1, 2, 3, 4, 5 lie on Layer II; 6, 7, 8, 9, 10, 11 on Layer III; etc.

In the circuit shown in Figure 7.6 each rosette is described only by the nearest neighbor inhibition ($\bullet$). The other connections shown in Figure 7.4 are not shown. All VCONs excite the GI, which explains the column of open circles at the right of this diagram, and the GI inhibits all of the other VCONs, which explains the row of dots along the bottom (GI output). The open circles on the sub and super diagonals indicate the excitatory projections between layers. Inputs to Layer IV are shown at the top under IV and into the GI. Outputs are shown emanating from the output lines (left) from VCONs in Layer IV and from the GI. Inputs and outputs can be conveniently displayed on this diagram by plotting the respective voltages on the lines putting into or coming out from respective VCONs.

7.2.3.1 Pencil model cortex

We next describe arrays of pencil columns. Visualize a two-dimensional array of columns much like a handful of pencils. We refer to the position of a pencil in this array as being its site. Suppose there are N_x sites in the x direction and N_y in the y direction. See Figure 7.7.

Let $\theta_{ijk}(t)$ denote the phase of VCON i in the pencil at site jk. Thus, j, k is the address of the column site and i is the address of the VCON within a column. Each pencil has 31 VCONs; $i = 30$ represents the Grand Inhibitor for the column, $i = 0, \ldots, 5$ are the VCONs on Layer II, and on Layer $L +$ II lie VCONs labeled $i = L * 6 + 0, \ldots, L * 6 + 5$ for $L = 0, \ldots, 4$.

$\{\theta_{ijk}(t)\}$ solves the system of equations

$$\dot{\theta}_{ijk} = 1 + \cos\theta_{ijk} + \sum_{l=0}^{30}\sum_{m=0}^{N_x}\sum_{n=0}^{N_y} B_{l,m,n,i,j,k}\cos_+\theta_{l,m,n} + \sum_{L=0}^{30} C_{L,i}\cos_+\theta_{L,j,k}$$

for $i \in \overline{0, 30}$, $j \in \overline{1, N_x}$, and $k \in \overline{1, N_y}$. Recall that N_x and N_y are the number of columns in the x and y directions, respectively.

If the site j, k in our simulations is forced with strength $\sigma_{i,j,k}$, then an additional $\sigma_{i,j,k}$ is added to the right-hand side. The matrix C describes connections within one column as described in Figure 7.6. The array B describes connections between columns. $B_{l,m,n,i,j,k}$ is zero unless i corresponds to Layer IV or to the GI. Otherwise, it indicates the strength of connection from the element at position l in the column at site m, n to the element at position i in the column at site j, k.

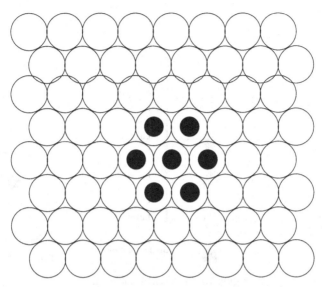

Figure 7.7. Depicted here is an 8 × 8 array of pencils. The Grand Inhibitors of 7 central ones are shown to highlight the hexagonal structure used here, which is similar to packing of cells in a plane.

7.2.4 Pencil stub model: a neuro-oscillator

The pencil model is quite complicated (though it is simple compared to a real cortical column). If we wish to consider large networks of cortical columns that are tractable for computer simulations, we might wish to further simplify the model.

First, we could replace all of the excitatory components in the column medulla by a single excitatory neuron. This gives the Exciter-Inhibitor model. Using a simple low-pass filter to model each neuron (E for excitation, I for inhibition), one arrives at the model

$$\dot{\theta} = a,$$

$$\tau_E \dot{a} = -a + f(\theta, a, \phi, b),$$

$$\dot{\phi} = b,$$

$$\tau_I \dot{b} = -b + g(\theta, a, \phi, b),$$

where f and g are chosen according to which model [16] one chooses to use. Such models have been used effectively to identify some interesting threshold properties of networks [140, 141].

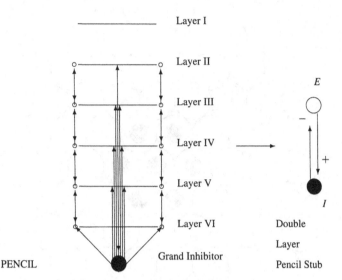

Figure 7.8. The pencil stub model.

An interesting model results if we replace all of the excitatory components in the column medulla by a single excitatory element with phase θ and consider it along with the GI with phase ϕ. This gives the VCON *neural oscillator model*

$$\tau_E \ddot{\theta} + \dot{\theta} = 1.0 + V(\theta) + C_{1,0} V_+(\phi),$$

$$\tau_I \ddot{\phi} + \dot{\phi} = 1.0 + V(\phi) + C_{0,1} V_+(\theta),$$

where $C_{1,0} < 0$ and $C_{0,1} > 0$. This is depicted in Figure 7.8 as the double layer model. Such double-layer structures have been used effectively to identify interesting threshold properties of networks [140].

Since the trajectories of this double layer oscillator usually describe simple clocks, we might replace it by a single VCON. The next section shows how this model enables us to perform some large-scale simulations of the visual cortex.

Our study resides in the frequency domain, so we continue with analysis of VCONs on each layer. The idea in reduction of complexity is to get similar but simpler models to facilitate analyzing large networks.

7.3 Ocular dominance

Innervation of the visual cortex can lead to interesting correlations between inputs and columns. Through learning synapses, these could drive the formation of striated connections similar to the *ocular dominance patterns* observed in cat visual cortex [70, 98, 109].

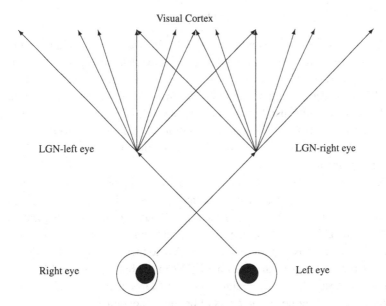

Figure 7.9. Organization of the visual system.

A newborn visual cortex is uniformly innervated with input from both eyes. Normal visual cortex development results in nonoverlapping patches, each with input connections that are dominated by one eye or the other. These patches are referred to as ocular dominance columns. Ocular dominance is a self-organizing property of the primary visual cortex in which inputs from both eyes pass through the lateral geniculate nucleus (LGN) of the thalamus (see Figure 7.9), and resulting in the alternating patches of connections. This self-organization may be important to cognitive processes such as stereopsis. Most stimuli that appear in the visual field are received by both eyes and then are relayed to the visual cortex on the side opposite to the visual field where the stimuli appear. The visual cortices of both hemispheres receive afferents from both eyes. Presented here is a VCON model that simulates normal development of ocular dominance columns.

Ocular dominance is believed to develop as a result of competition between the inputs of the two eyes [98]. In addition, inputs from the LGN are progressively pruned, further refining the patches [98].

Some synapses are believed to vary in strength according to Hebbian learning mechanisms that essentially correlate the activity of the neurons on either side of the synapse [47]. A high degree of correlation between neurons will cause the synapse to strengthen and low correlation allows the synapse to weaken.

7.3.1 The model

The network used in simulation here is a 26 by 26 array representing a section of the primary visual cortex. This array receives overlapping input from a pair of 26 by 26 arrays of input nodes representing two laminae of the LGN. One LGN lamina is dedicated to the left eye, and the other is dedicated to the right eye. Each cortical site also has lateral connections to other cortical sites. The ocular dominance columns of interest are ultimately exhibited in the pattern of synaptic strengths between the cortical sites and the LGN nodes from which they receive input.

The visual cortex is modeled here using a 26 by 26 array of VCONs, where each VCON represents one cortical column. Each row of VCONs is staggered with respect to adjacent rows so that each VCON has a hexagonal configuration of nearest neighbors. Hence, there is symmetry under rotation by 60° (see Figure 7.7). The array is periodically extended so that VCONs in the top row are viewed as being contiguous with VCONs in the bottom row, and VCONs in the right-most column are contiguous with VCONs in the left-most column. Periodic extension removes the effect of boundary conditions.

Each cortical VCON has lateral connections with its neighbors some of which are excitatory and some inhibitory. These connections are symmetric and have the same magnitude and sign for all cortical VCONs of equal distance. Two cortical VCONs, say v and w, are considered to be of equal distance from a cortical VCON, say u, if they both lie on the perimeter of a hexagon of which u is the center.

In the following discussion and results, the lateral connections are taken to be in "Mexican Hat" configuration in which proximal neighbors make excitatory connections, and more distant VCONs make inhibitory connections.

It is well known that the LGN, like the primary visual cortex, is retinotopically organized and that combined processing of input from both eyes does not occur in the LGN [81]. The activity of the LGN is therefore taken to be a sufficient representation of input to the primary visual cortex since this activity is a mirror image of retinal activity.

The LGN is modeled like the visual cortex as being a pair of 26 by 26 arrays of input nodes. Each input node represents one neuron in the LGN. Each array of the LGN is configured identically to the cortex, but one array represents input from the left eye and the other input from the right eye. In the real visual system, both arrays receive overlapping input from the same half of the visual field, but from different eyes.

Each cortical VCON receives input from the corresponding nodes in each of the two LGN arrays and from the set of nearby neighbors to them.

The LGN arrays serve as the input layers for the cortical VCONs. We suppose that each input node has a phase

$$Y_i = \omega_i t,$$

where ω_i is the firing rate of the LGN neuron represented by the input node, and t is the current time.

The firing rates, ω_i, of the input nodes in one of the two LGN arrays are set to random numbers from 0.0 to 6.0. The firing rates of the input nodes in the other array are set to the same values as the first array but are shifted along one spatial dimension by some fixed number of sites. This shift factor is referred to as the ocular spatial phase shift and is denoted Δx.

The mathematical model comprises a system of three arrays of differential equations. Equation (7.5) describes the phase of each cortical VCON, $j = 1, \ldots, 26^2$. The other two arrays of Equations (7.6) and (7.7) describe the synaptic strengths of the connections between the input nodes, i, of the lateral geniculate nucleus and the VCONs, j, of the visual cortex. Each cortical VCON receives connections from 7 input nodes in the LGN, yielding 676×7 equations for each array (7.6) and (7.7). Thus the entire system comprises $2 \times 676 \times 7 + 676 = 10, 140$ equations that must be solved.

The resulting model for development of the visual cortex is

$$\dot{X}_j = 1.0 + \left(\sum_{i=1}^{N} \left[C_{ij}^L V(Y_i^L) + C_{ij}^R V(Y_i^R) \right] + \sum_{k=1}^{N} I(k, j) V(X_k) \right) V(X_j),$$
$$(7.5)$$

$$t\dot{C}_{ij}^L = -C_{ij}^L + G V(Y_i^L) V(X_j),$$
$$(7.6)$$

$$t\dot{C}_{ij}^R = -C_{ij}^R + G V(Y_i^R) V(X_j),$$
$$(7.7)$$

for each $j \in \overline{1, 676}$ and for each i from which the jth site receives input. Here

$V(\omega t) = \cos_+^4(\omega t)$ (fourth power of rectified cosine),

$t = $ time,

$G = $ gain factor representing the learning rate,

$X_j = $ phase of cortical VCON at site j,

$C_{ij}^L, C_{ij}^R = $ synaptic strength of connection from LGN input node at site i (arrays corresponding to left (L) and right (R) eyes, respectively) to cortical VCON at site j,

$Y_i^L, Y_i^R = $ phase of LGN input node at site i (arrays corresponding to left and right eyes, respectively) (recall that these are $Y_i = \omega_i t$), and

$I(k, j) = $ strength of lateral connection between cortical VCONs k and j.

The terms $V(Y_i^L)$ and $V(X_j)$ represent the signal of the presynaptic and postsynaptic nodes (LGN input nodes and cortical VCONs), respectively. The activity of the nodes is correlated by integrating the product. A large degree of correlation results in faster synaptic growth.

Equations (7.6) and (7.7) can be rewritten as

$$C_{ij}^{L,R} = \frac{G}{t} \int_0^t V\left(\omega_i^{L,R} s\right) V\left(X_j(s)\right) ds$$

for each VCON j and all i from which it receives input. This representation emphasizes that these are correlations.

The initial strengths for all synapses are set to zero in the simulation to illustrate that initial synapse strength plays little or no role in causing the characteristic patterns of ocular dominance. Similarly, the initial phases of all cortical VCONs have been set to unity to show that a relative difference in the initial level of activity in cortical VCONs is not needed to produce ocular dominance patterns.

7.3.2 Model outputs

Output for this model is generated for each cortical VCON. The sum of the synaptic strengths of the connections from the right LGN array is subtracted from the sum of the synaptic strengths of the connections from the left LGN lamina. For visualization purposes, we use as a measure fifty times the hyperbolic tangent of this difference. This results in a measure proportional to left eye dominance. This measure is converted to a gray level (0 to 255) so that black pixels (gray level = 0) correspond to right eye dominance and white pixels (gray level = 255) to left eye dominance.

The results are similar to those obtained by staining techniques used in biological studies: We use a threshold parameter A; all measures above $+A$ are set to gray level 255 (full left eye dominance); all measures below $-A$ are set to gray level 0 (full right eye dominance). All measures between $-A$ and $+A$ are set to a linearly interpolated gray level. The value of A used is 45.0, which is roughly 90% of the range observed in simulations.

Results are presented in a 40 × 40 array comprising the original 26 × 26 array and its periodic extension. Using 40 rows and columns allows the patterns occurring at the boundaries to be more easily seen and is consistent with the periodic extension connection scheme described earlier. Each VCON is given scale by an 8 by 8 block of pixels so that the entire image has dimensions of 320 by 320 pixels. Figure 7.10 summarizes the output of this simulation.

Ocular dominance patterns emerge from the patternless primary visual cortex, and they gradually become more evident with time. The surface of

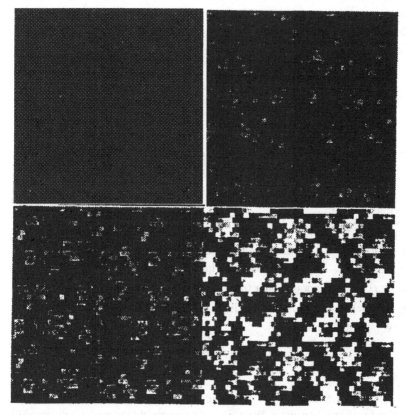

Figure 7.10. Simulated ocular dominance pattern development at times $t = 0$ (upper left), 1 (upper right), 4 (lower left), and 7 (lower right). White denotes left eye dominance, and black denotes right eye dominance. The simulation begins with a salt-and-pepper distribution of connections, and ocular dominance patches emerge. (Redrawn with permission from [109].)

the cortex is almost completely segregated into patches subserving one eye or the other by the end of the critical period, except for a small number of binocular cells at the borders between patches. Although no two patches are exactly the same size, there does tend to be a rough spatial periodicity to the patterns that develop. Large patches correspond to a longer spatial period (or smaller spatial frequency).

7.4 Nonlinear waves in a continuum model

In this section, we derive continuous models of networks and analyze them for wave propagation features. Consider a continuum of VCONs that are coupled by some connection scheme. Let $\boldsymbol{\theta}(s, t) \in E^N$ denote the phases of circuit elements at location $s \in E^3$ and $\mathbf{y}(s, t) \in E^M$ the vector of phases of inputs. Then we have

$$\frac{\partial \boldsymbol{\theta}}{\partial t} = \mathbf{a},$$

$$\tau(s)\frac{\partial \mathbf{a}}{\partial t} = -\mathbf{a}(s, t) + \boldsymbol{\omega}(s) + \cos \boldsymbol{\theta}(s, t)$$

$$+ P\left(\mathbf{x}(s, t), \iiint_{-\infty}^{\infty} \left\{ K(s, s')V[\boldsymbol{\theta}(s', t)] \right.\right.$$

$$\left.\left. + L(s, s')V[\mathbf{y}(s', t)] \right\} ds' \right).$$

For these equations:

- At location $s \in E^3$ there may be many circuits having different phases, say $\{\boldsymbol{\theta}_1(s, t), \ldots, \boldsymbol{\theta}_N(s, t)\}$, and many inputs, say $\{y_1(s, t), \ldots, y_M(s, t)\}$.
- $\boldsymbol{\omega}(s)$ is the vector of center frequencies in the network.
- $\mathbf{a}$ is the vector of activities as discussed in Chapter 3.
- We use the notation

$$\cos \boldsymbol{\theta} = (\cos \boldsymbol{\theta}_1, \ldots, \cos \boldsymbol{\theta}_N),$$

etc.

- The kernel $K(s, s') \in E^{N \times N}$ denotes the strength of connections from site (or location) s' to site s. We consider here only the case where the strength of connection depends solely on distance between the sites, so we write

$$K(s, s') = K(s - s').$$

Similarly, the polarity of connections is $\psi = \psi(s - s')$.

- The kernel $L(s, s') \in E^{N \times M}$ denotes the strength of connection and polarity from input site s' to site s. As above, we consider here only the case where

$$L(s, s') = L(s - s').$$

- All data and their derivatives are assumed to be continuous functions.
- The function $P : E^N \times E^N \to E^N$ can be chosen in a variety of ways as discussed earlier.

Note that all of the synapses are ignored in that we do not account for synaptic time delays or learning. In addition, this model does not account for connections through activities. The model can be further complicated by including chemical synapses and synaptic filtration, but neither is incorporated here.

Because we expect only quasi-constant forcing terms to survive the low-pass filter in this system, we can consider a plausible associated problem that could be derived through Fourier analysis and perturbation analysis of this system. Consider the equation

$$\tau \frac{\partial^2 \theta}{\partial t^2} + \frac{\partial \theta}{\partial t} - \cos \theta$$

$$= \omega + \int_{-\infty}^{\infty} \int_{-\infty}^{\infty} \int_{-\infty}^{\infty} K(s - s') f\left[\theta(s, t) - \theta(s', t) - \psi(s - s')\right] ds',$$

where the kernel K, described above, gives the amplitude of connection from site s' to site s and $\psi(s - s')$ gives the polarity of the connection from site s' to site s [67]. f is a smooth function.

Such equations were studied for $s \in E^1$ by Garcia in [35]. She showed that if $\psi \equiv 0$ this equation has stable *bulk oscillations* of the form $\theta(s, t) \approx \omega t + \phi_0$. This implies uniform synchrony in the network.

We will avoid technical mathematical difficulties by restricting attention to the case where K and ψ have period L and f has period 2π in each component of θ. With these assumptions, we consider the equation

$$\tau \frac{\partial^2 \theta}{\partial t^2} + \frac{\partial \theta}{\partial t} - \cos \theta$$

$$= \omega + \int_{-L}^{L} K(s - s') f\left[\theta(s, t) - \theta(s', t) - \psi(s - s')\right] ds'.$$

We have written a single integral although it might be a triple integral if $s \in E^3$, a double integral if $s \in E^2$, etc.

There are also steady progressing wave solutions of these equations in one, two, and three dimensions, and we discuss these next.

7.4.1 Steady progressing waves in one-layer

A constant theme throughout this work is that conversion of a problem to the frequency domain can greatly simplify some important calculations. This is also the case in seeking nonlinear wave solutions of network models.

Consider the equation

$$\tau \ddot{\theta} + \dot{\theta} = \omega + \int_{-L}^{L} K(s - s') f\left[\theta(s, t) - \theta(s', t) - \psi(s - s')\right] ds',$$

where τ, ω, K, and ψ are given.

It is not obvious how a steady progressing wave should be defined for this problem. By a *steady progressing wave solution* we mean here a solution to this problem of the form

$$\theta = ks + ct,$$

where the number k gives the spatial wave number and c gives the propagation speed. Such a solution would be observed through a periodic function $V(ks+ct)$. If this were plotted, the solution would look like a wave having period $2\pi/k$ in the s direction travelling at velocity c to the left (if $c > 0$). Substituting this form into the equation gives

$$c = \omega + \int_{-L}^{L} K(s - s')f\left[k(s - s') - \psi(s - s')\right]ds',$$

which determines a unique wave speed c for the wave as a function of wave number k *if* the integral is constant. This requires that $Lk = 2\pi m$ for some integer m. When this condition is satisfied, there is a corresponding wave solution. Thus, there is a rich structure of many waves having various spatial and temporal frequencies.

The stability of these waves can be determined in the following way: We consider here the case where $L = \pi$ and $k = 1$, and we define

$$\phi = \theta - s - ct.$$

There results an equation for ϕ:

$$\begin{aligned}
\tau\ddot{\phi} + \dot{\phi} &= \int_{-\pi}^{\pi} K(s - s')\Big\{ f\left[\theta(s, t) - \theta(s', t) - \psi(s - s')\right] \\
&\quad - f\left[s - s' - \psi(s - s')\right]\Big\}ds' \\
&= \int_{-\pi}^{\pi} K(s - s')\Big\{ f'\left[s - s' - \psi(s - s')\right][\phi(s, t) - \phi(s', t)]\Big\}ds' \\
&\quad + O\Big\{|[\phi(s, t) - \phi(s', t)]|^2\Big\}.
\end{aligned}$$

Since the data are smooth and periodic, they have convergent Fourier series expansions. Therefore, we seek the solution of the linear problem in the form

$$\phi(s, t) = \sum_{n=-\infty}^{\infty} \phi_n(t)e^{ins}$$

Then

$$\tau\ddot{\phi}_n + \dot{\phi}_n = (A_0 - A_n)\phi_n,$$

where $\{A_n\}$ are the Fourier coefficients of the kernel

$$K(s)f'(s - \psi(s)) = \sum_{n=-\infty}^{\infty} A_n e^{ins}.$$

The stability of the various modes of ϕ depend on the coefficients $\{A_n\}$. If $K(s) = f(s - \psi(s)) = \cos s$, then some modes oscillate in time. If $f(s - \psi(s)) = \sin s$, then all modes with $n \neq 0$ decay exponentially.

7.4.2 Steady progressing waves in two-layers

The same analysis can be carried out for a double-layer system: Consider the network

$$\tau_1 \ddot{\theta}_1 + \dot{\theta}_1 = \omega_1 + \int_{-L}^{L} K_{11}(s - s')f[\theta_1(s, t) - \theta_1(s', t) - \psi_{11}(s - s')]\,ds'$$

$$+ \int_{-L}^{L} K_{21}(s - s')g[\theta_1(s, t) - \theta_2(s', t) - \psi_{21}(s - s')]\,ds',$$

$$\tau_2 \ddot{\theta}_2 + \dot{\theta}_2 = \omega_2 + \int_{-L}^{L} K_{22}(s - s')F[\theta_2(s, t) - \theta_2(s', t) - \psi_{22}(s - s')]\,ds',$$

$$+ \int_{-L}^{L} K_{12}(s - s')G[\theta_2(s, t) - \theta_1(s', t) - \psi_{12}(s - s')]\,ds'.$$

We seek a solution of the form

$$\theta_1(s, t) = s + c_1 t, \quad \theta_2(s, t) = s + c_2 t;$$

substituting this into the system gives

$$c_1 = \omega_1 + \int_{-L}^{L} K_{11}(s - s')f[s - s' - \psi_{11}(s - s')]\,ds'$$

$$+ \int_{-L}^{L} K_{21}(s - s')g[s - s' - \psi_{21}(s - s') + (c_1 - c_2)t)]\,ds',$$

$$c_2 = \omega_2 + \int_{-L}^{L} K_{22}(s - s')F[s - s' - \psi_{22}(s - s')]\,ds'$$

$$+ \int_{-L}^{L} K_{12}(s - s')G[s - s' - \psi_{12}(s - s') - (c_1 - c_2)t)]\,ds'.$$

The center frequencies can be quite different; for example, in the thalamus-reticular complex model, their ratio is approximately 100:1. If we take $c_1 \approx \omega_1$ and $c_2 \approx \omega_2$ and suppose that the ratio of the two is large, then the second integrals on the right-hand side are highly oscillatory and will not contribute to the response through the low-pass filter.

The result with these assumptions is that there are two waves, one on the top and one on the bottom. They would be observed as

$$V(s + \omega_1 t)$$

and

$$V(s + \omega_2 t),$$

respectively. These waves have the same wave number in space, but they have very different propagation speeds: The one on the top moves 100 times faster than the one on the bottom.

7.4.3 Tracking waves of input

A layer of excitable cells can track an input. For example, consider a two-dimensional array of VCONs modeled by

$$\dot{\theta} = \omega(s, t) + \cos \theta(s, t),$$

where $s \in R^2$ labels the sites. Input to this layer is accounted for in the function ω. For example, if

$$\omega(s, t) = A \operatorname{sech} (\eta_1 s_1 + ct) \operatorname{sech} (\eta_2 s_2 + ct),$$

where $A > 1$, then as t increases, a bump of excitation is observed to move along the straight line having direction numbers (η_1, η_2) with speed $- c$. In response, the sites will fire where $\omega > 1$ and a bump of bursting activity will be observed in the network.

Presumably, adding a network of connections to this array will result in a bump of activity moving as a steady progressing wave with velocity c in the direction η. Some sort of inertia should result in a bump continuing for sometime after we set $A = 0$. This would suggest that the network anticipates motion of the input [121].

7.4.4 Numerical simulation of waves

Computer simulations of continuum models require some sort of conversion to a discrete structure that a computer can process. In this section, we will approximate the continuum network by a discrete system that is suitable for numerical simulation of the network. This is done by approximating a function

$\theta(s, t)$ for $0 \leq s \leq S$ by a vector $\mathbf{x}(t) = (x_1(t), \ldots, x_N(t))$, where

$$x_j(t) = \theta(jS/N, t)$$

for $j = 1, \ldots, N$.

The rest of this chapter is devoted to deriving and analyzing equations for the vector $\mathbf{x}(t)$ in various situations.

We first deal with discrete approximations to the convolution integrals appearing in the continuum model.

7.4.4.1 Reduction of convolution integrals

The convolution integrals that appear in the models above can be replaced by a simple quadrature formula. We begin by considering a general convolution integral. Suppose that $f(s)$ is an S-periodic function (i.e., $f(s + S) \equiv f(s)$ for all s), and let us define a function $g(s)$ by the convolution integral

$$g(s) = \int_{-\infty}^{\infty} K(s')f(s - s')\,ds'.$$

If K has finite support (say $K = 0$ for $|s| \geq S$), we have that

$$\begin{aligned}
g(s) &= \int_{-S}^{S} K(s')f(s - s')\,ds' \\
&= \int_{0}^{S} K(s')f(s - s')\,ds' + \int_{-S}^{0} K(s')f(s - s')\,ds' \\
&= \int_{0}^{S} K(s')f(s - s')\,ds' + \int_{0}^{S} K(u' - S)f(s - u' + S)\,du' \\
&= \int_{0}^{S} \left[K(s') + K(s' - S)\right]f(s - s')\,ds'.
\end{aligned}$$

Now, we define

$$s_j = jS/N$$

for $j = 1, \ldots, N$, and we replace the integrals by sums that approximate them using lower rectangles:

$$\begin{aligned}
g(s_m) &= \frac{S}{N} \sum_{j=1}^{N} \left\{ K(jS/N) + K[(j - N)S/N)]f[(m - j)S/N] \right\} \\
&= \sum_{j=1}^{N} k_j f_{m-j},
\end{aligned}$$

where

$$k_j = \frac{S}{N}\left\{K(jS/N) + K\big[(j-N)S/N\big]\right\}$$

and

$$f_{m-j} = f\big[(m-j)S/N\big]$$

for $j = 1, \ldots, N$. Setting $g_m = g(s_m)$, we have that

$$g_m = k_{m-1}f_1 + \cdots + k_1 f_{m-1} + k_N f_m + \cdots + k_m f_N.$$

Writing this as a vector problem, we have that

$$\mathbf{g} = C\mathbf{f},$$

where

$$\mathbf{g} = \begin{pmatrix} g_1 \\ \vdots \\ g_N \end{pmatrix}, \quad \mathbf{f} = \begin{pmatrix} f_1 \\ \vdots \\ f_N \end{pmatrix},$$

and the matrix C is given by

$$\begin{pmatrix} k_N & k_{N-1} & \cdots & k_1 \\ k_1 & k_N & \cdots & k_2 \\ \vdots & \vdots & \ddots & \vdots \\ k_{N-1} & \cdots & k_1 & k_N \end{pmatrix}.$$

The matrix C is called a *cyclic matrix* because of its special structure.

Note that if in addition $K(s) = K(-s)$ (the kernel is an even function of s), then

$$k_N = (S/N)\big[K(S) + K(0)\big],$$

$$k_{N-1} = k_1 = (S/N)\left\{K(S/N) + K\big[(N-1)S/N\big]\right\}, \text{ etc.}$$

Therefore, the matrix C is a real symmetric matrix. The convolution integral can be visualized as shown in Figure 7.4 with appropriate weights on connections. Simulation and presentation of the results is described in the exercises.

7.4.5 Analysis of a cyclic matrix

Cyclic matrices are attractive because all of their eigenvalues and eigenvectors can be easily found (see [25]). Let ρ be the generator of the Nth roots of unity:

$$\rho = \exp\left(\frac{2\pi i}{N}\right).$$

Then the jth eigenvalue of C is given by

$$\lambda_j = \sum_{m=1}^{N} k_{N-m} \rho^{mj}$$

and the corresponding right eigenvector is given by

$$\phi_j = (\rho^j, \ldots, \rho^{j(N-1)}, \rho^{Nj}).$$

These are complicated expressions, but the fact that they are given in closed form is very useful. In particular, they can be computed easily. Note that since C is a real symmetric matrix, the eigenvectors corresponding to different eigenvalues are orthogonal. We will make use of four cases in later examples: $N = 2, 3, 4$, and 8.

For $N = 2$, we have that

$$C = \begin{pmatrix} k_2 & k_1 \\ k_1 & k_2 \end{pmatrix}.$$

In this case, $\rho = -1$ and the eigenvalues and eigenvectors of C are

$$\lambda_1 = k_2 - k_1,$$

$$\phi_1 = (-1, 1)$$

and

$$\lambda_2 = k_1 + k_2,$$

$$\phi_2 = (1, 1).$$

Therefore, we can write C in its spectral decomposition as

$$C = \frac{\lambda_1}{\sqrt{2}} \begin{pmatrix} 1 & -1 \\ -1 & 1 \end{pmatrix} + \frac{\lambda_2}{\sqrt{2}} \begin{pmatrix} 1 & 1 \\ 1 & 1 \end{pmatrix}.$$

This form is useful for computations since it converts matrix multiplication to scalar multiplication; for example,

$$C^2 = \frac{\lambda_1^2}{\sqrt{2}} \begin{pmatrix} 1 & -1 \\ -1 & 1 \end{pmatrix} + \frac{\lambda_2^2}{\sqrt{2}} \begin{pmatrix} 1 & 1 \\ 1 & 1 \end{pmatrix},$$

etc. Other useful cases are where $N = 3, 4$, or 8 (See [57]).

7.4.6 Rosettes

We now return to the derivation of the rosette structure used on medullar layers in the column model, but now from the perspective of the continuum model using the reduction via cyclic matrices.

Recall that the vector **x** describes the phases at the network's sites and that our model is for this variable. We begin by quantizing the convolution integral as described earlier. We write

$$\int_{-\infty}^{\infty} K(s - s') f(s')\, ds' \equiv Cf,$$

where C is a cyclic matrix derived by approximating the integrals by matrix multiplications acting on a discretization of **f**. With this notation, the model becomes, as in Equation (7.4),

$$\dot{x}_j = \omega_j + \cos x_j + V(x_j) + \sum_{k=1}^{N} C_{j,k} V_+(x_k)$$

for $j = 1, \ldots, N$.

We refer to this as a neuron rosette, since the VCONs can be visualized as being distributed around a circle with external stimuli being arrayed around a concentric circle. The connections between the neurons are described by the components of C, where $C_{k,j}$ denotes the strength of connection from neuron k to neuron j.

Although this model appears to be quite complicated, it can be studied rigorously using the rotation vector method if $1/|\omega| \ll 1$. To do this, we first set the data in the problem. These are C and ω. Once these are set, the Fourier expansion for the right-hand sides can be determined numerically using fast Fourier transforms methods. The rotation vector method can then be used to discover the phase-locking properties of the network. However, since we know to expect either a fixed pattern of phase values or phase locking, we can proceed directly with interesting numerical simulations of the rosette.

7.5 Summary

This book was motivated by four events: 1. the discovery of a convenient method for phase-locking analysis [60]; 2. the construction of a VCON model that facilitates frequency-domain analysis, or signal processing, of networks; 3. various work on Ising-like models to study large networks; and 4. observations by R. Guttman and others that squid axons have frequency and phase deviation relations between inputs and outputs and have rich phase-locking properties. We added to this some facts that have been obtained by neuroscientists more recently.

Our discussion of large networks was split into two parts. In Chapter 6, we studied mnemonic surface approaches to them and described some analytic and discrete methods for studying them. In Chapter 7, we studied problems arising in known brain networks using some of the methods derived in Chapter 6 and earlier.

Our approach has led to a number of results:

- Models can be formulated and studied in the frequency domain that are useful in understanding the formation and processing of signals in oscillatory networks.

- The models are quite general in the sense that any model formulated in terms of dynamical systems and that involves certain simple bifurcation phenomena can be reduced to the VCON model.

- Signaling in the brain involves both the firing frequency of signals and their timing, or phase deviations. Our model neurons are comparable to FM radios.

- Networks can memorize phase deviations. This is studied in greater detail in [75].

- There are rich structures of wave propagation in VCON networks. We have started analysis of these phenomena by considering the existence of steady progressing waves in them.

Many important questions are not addressed in this book. For example, precisely what effect does an unstable connection kernel have on spatial patterns of phase locking and phase deviations? What about unstable lateral stimulation that is described between layers? What is the effect of nonuniform center frequencies, say where one region should phase lock at frequencies different from those of another? Different neurons phase locking onto different frequencies will identify communicating groups within a network by self-organization. What is the correct analogue of mnemonic surfaces in networks with hysteretic coupling, e.g., through gate synapses? In general, one expects the mnemonic surface to not be single valued.

This book presents some evidence that phase locking between model neurons can be a mechanism of memory; this mechanism is consistent with some intuition about our own neural networks. The phase-locking patterns described here depend on the connections between neurons in the network, and so they result from the network's structure. Therefore, the patterns should reestablish themselves after interruption of electrical activity by electrical shock or anesthesia.

As mentioned at the start of this chapter, the focus in this book has been on mathematical methods for the neurosciences, and we have studied examples here to illustrate approaches to modeling brain structures and methods for mathematical analysis of them. Mathematics is needed to understand how the brain works, and future work will build on the mathematical techniques derived for models from the neurosciences and from work described here. In particular, analysis of canonical models in the frequency domain is essential for understanding the flow of information in networks, in particular in networks of neurons and the brain.

7.6 Exercises

1. *Cyclic matrix.* The reduction of a convolution integral in Section 7.4.6 results in the formula

$$\mathbf{g} = C\mathbf{f}$$

where C is a special cyclic matrix. Derive the coefficients of C.

2. *Spectral decomposition of C.* Derive the spectral decomposition of the cyclic matrices C in Section 7.4.6.

3. *Pencil stub.* In the pencil model of a cortical column, replace the column medulla by a single excitatory VCON. Derive the model corresponding to this simplification. Describe possible dynamics of such a pair (the VCON representing the medulla and one representing the Grand Inhibitor). Note that the atoll model describes such a pair.

4. Consider an 8×8 array of cortical columns as modeled in the preceding exercise. Carry out several computer simulations of this network assuming first that there are no connections between the columns and all receive a fixed (constant) stimulation in the excitatory VCON, and second that there are only excitatory connections from the medulla to the grand inhibitor of neighbors.

5. Carry out the wave propagation analysis in Section 7.4 for the cases where (a) $f(\theta) = \cos\theta$ and $K(s) = \cos s$, and (b) $f(\theta) = \sin\theta$ and $K(s) = \cos s$. Derive formulas for all possible wave speeds in these two cases.

6. Show that if a pulse of activity is given at $s = 0$ and later one at $s = \pi$, then the two waves generated by these might cross. Suppose as in the study of sound location, there is no response of a monitoring system until activity is of sufficient size. Show that this system is capable of leaving a residual pattern from such transient events as interacting steady progressing waves. Carry out computer simulations of the models to demonstrate your point.

Appendix A

Mathematical background

The material in this section describes some mathematical methods and models used in studying biological rhythms. Most of these topics involve differential equations related to electrical circuits, and the following is a brief introduction to how such equations are solved.

A.1 Examples

A.1.1 Low pass filter

Let $x(t)$ denote the capacitor voltage at time t. Kirchhoff's laws suggest that this voltage changes according to the equation:

$$dx/dt = -\alpha x, \quad x(0) = A.$$

Here α is a constant called the decay rate and A is the initial voltage.

This equation can be solved by the method of separation of variables: We write

$$dx/x = -\alpha dt,$$

$$\log(x/A) = -\alpha t,$$

$$x(t) = A \exp(-\alpha t).$$

The voltage becomes roughly half its value every $1/\alpha$ time units, so $1/\alpha$ is called the time constant of the circuit.

A.1.2 Harmonic oscillator

A basic model in physics is the harmonic oscillator,

$$\ddot{x} + \omega^2 x = 0,$$

where $\dot{x} = dx/dt$ and $\ddot{x} = d^2x/dt^2$. Here ω is a known constant called the free frequency, and we specify initial conditions

$$x(0) = A, \quad \dot{x}(0) = B,$$

where A and B are known.

The harmonic oscillator equation cannot be solved using the model of separation of variables, so a more sophisticated approach is taken that results in the following fact: Given any two linearly independent solutions, say $x_1(t)$ and $x_2(t)$, *any* solution of the harmonic oscillator can be written as a linear combination of them. That is, the form

$$x(t) = a_1 x_1(t) + a_2 x_2(t),$$

where a_1 and a_2 are free constants, covers all possible solutions.

The functions $x_1(t) = \cos \omega t$ and $x_2(t) = \sin \omega t$ each satisfy the harmonic oscillator equation. They are linearly independent, since the only linear combination of them that vanishes identically is the trivial one:

$$c_1 \cos \omega t + c_2 \sin \omega t \equiv 0$$

for all t implies that when $t = 0$, $c_1 = 0$, and when $t = \pi/(2\omega)$, $c_2 = 0$. Therefore, $c_1 = c_2 = 0$.

It follows from the fact above that any solution of the harmonic oscillator equation can be written in the form

$$x(t) = a_1 \cos \omega t + a_2 \sin \omega t.$$

The choice $a_1 = A$ and $a_2 = B/\omega$ picks out the unique solution that satisfies both of the initial conditions $x(0) = A$ and $\dot{x}(0) = B$.

Although this method was fairly straightforward, another approach, the Laplace transform method, also proves useful in enabling us to find linearly independent solutions.

The Laplace transform method (short version) proceeds in the following way: Let $x(t) = \exp(rt)$ and try to find what r must be so that x is a solution of the equation. Putting this solution into the equation, we get

$$(r^2 + \omega^2) \exp(rt) = 0.$$

The two possible choices are $r = i\omega$ and $r = -i\omega$, where i is the square root of -1. Therefore, two solutions of the equation are

$$x_1(t) = \exp(i\omega t), \quad \text{and} \quad x_2(t) = \exp(-i\omega t)$$

(see Exercises 1).

A.2 Forced harmonic oscillator

An externally forced harmonic oscillator is described by the equation

$$\ddot{x} + \omega^2 x = f(t),$$

where $f(t)$ is a known external forcing function. This problem can be solved by the method of variation of parameters. We first make an educated guess that the solution has the form

$$x(t) = a(t)\cos\omega t + b(t)\sin\omega t,$$

where $a(t)$ and $b(t)$ must be determined. Whether or not this solution works justifies or invalidates the guess. We have two unknowns to find, so we try to obtain two equations for them. We arbitrarily set

$$\dot{a}\cos\omega t + \dot{b}\sin\omega t = 0.$$

Then we substitute the guess into the equation and see what happens. The result is that

$$-\dot{a}\omega\sin\omega t + \dot{b}\omega\cos\omega t = f(t).$$

Solving these equations for $\dot{a}$ and $\dot{b}$, we get

$$\dot{a} = -f(t)\sin\omega t/\omega, \quad \dot{b} = f(t)\cos\omega t/\omega.$$

Solving these for a and b, by integrating the equations, we have

$$a(t) = a(0) - \frac{1}{\omega}\int_0^t f(s)\sin\omega s\,ds$$

and

$$b(t) = b(0) + \frac{1}{\omega}\int_0^t f(s)\cos\omega s\,ds.$$

The final result is

$$x(t) = a(0)\cos\omega t + b(0)\sin\omega t - \frac{1}{\omega}\cos\omega t\int_0^t f(s)\sin\omega s\,ds$$

$$+ \frac{1}{\omega}\sin\omega t\int_0^t f(s)\cos\omega s\,ds.$$

Here $a(0)$ and $b(0)$ are free constants.

As an example, let $f(t) = A \cos vt$, where $\omega^2 - v^2 \neq 0$. In this case, we have from the formula that

$$x(t) = \left[a(0) - \frac{1}{\omega} \int_0^t f(s) \sin \omega s \, ds \right] \cos \omega t$$

$$+ \left[b(0) + \frac{1}{\omega} \int_0^t f(s) \cos \omega s \, ds \right] \sin \omega t$$

$$= x_0(t) + x_p(t),$$

where

$$x_0(t) = a(0) \cos \omega t + b(0) \sin \omega t$$

and

$$x_p(t) = A \cos vt / (\omega^2 - v^2)$$

if $\omega^2 - v^2 \neq 0$. The solution x_0 is called a general solution of the free problem, and x_p is a particular solution of the forced equation.

A.2.1 Resonance

If $\omega \approx v$ in the preceding example, then the solution is bounded (even oscillatory), but if $\omega = v$, then $x(t) \to \infty$. That is, if energy is fed into the system at the natural frequency ω, it keeps accumulating, but at any other frequency, no matter how exquisitely close to ω, it dissipates.

When $\omega = v$, we have

$$\ddot{x} + \omega^2 x = A \cos \omega t.$$

We see (by substitution) that

$$x(t) = \frac{At \sin \omega t}{2\omega}$$

is a solution of the equation.

A.2.2 Damped harmonic oscillator

The equation

$$\ddot{x} + R\dot{x} + \omega^2 x = 0$$

is referred to as the damped harmonic oscillator, and R is called the damping coefficient. The solution of this equation can also be found by the Laplace transform method: Setting $x(t) = \exp(rt)$ gives

$$r^2 + Rr + \omega^2 = 0.$$

The solutions of this equation are given by the quadratic formula

$$r = \frac{-R \pm \sqrt{R^2 - 4\omega^2}}{2}.$$

This result shows that if $R > 0$, then the real part of r is negative. It follows from this that $x(t) \to 0$ as $t \to \infty$.

A.2.3 Fourier methods

The equations that have been solved in the previous sections arise as models of a wide variety of physical and biological systems. Their usefulness has been demonstrated repeatedly by the experiments they can suggest and the explanations and detailed predictions they provide. Once a formula for the solution is found the job is still not complete since the formula must be evaluated in some useful way. This is not always easy.

Most of the functions to be evaluated here are oscillatory, and a good approach to such functions is provided by Fourier methods, i.e., Fourier series decompositions: It is known that a smooth oscillatory (periodic or almost periodic) function $f(t)$ can be expanded in terms of a Fourier series

$$f(t) = \sum_{n=0}^{\infty} c_n \exp(i\omega_n t),$$

where the amplitudes $\{c_n\}$ and the frequencies $\{\omega_n\}$ are determined by Fourier transform methods. This expansion is called a spectral decomposition of f because it breaks f down into terms that have specific frequencies. ω_n is the nth frequency, $\exp(i\omega_n t)$ is the nth mode of oscillation, and c_n is its amplitude. The ω_n are increasing in size, so successive terms have successively higher frequencies. Convergence of this series ensures that the amplitudes of higher modes are eventually smaller.

Such expansion procedures form the basis of signal processing and studies of circuits in the frequency domain. Since the equations considered to this point are linear in the unknown variable, the solution can be decomposed into Fourier series of which each terms solves an equation. Specifically, suppose that f is given by this series where its frequencies satisfy

$$|\omega_0| < |\omega_1| < \cdots \quad \text{and} \quad \lim_{n \to \infty} |\omega_n| = \infty$$

and its amplitudes are known. Then we can solve the forced harmonic oscillator using Fourier's method: Consider the equation

$$\ddot{x} + v^2 x = f(t).$$

Since f has a Fourier series expansion, let us suppose that x does also:

$$x(t) = \sum_{n=0}^{\infty} x_n \exp(i\omega_n t).$$

Substituting this into the equation and equating coefficients of like trigonometric terms, we have

$$(-\omega_n^2 + v^2)x_n = c_n.$$

Therefore,

$$x_n = c_n/(v^2 - \omega_n^2)$$

and

$$x(t) = \sum_{n=0}^{\infty} \frac{c_n}{v^2 - \omega_n^2} \exp(i\omega_n t)$$

If $v^2 \neq \omega_n^2$ for all $n = 0, 1, 2, \ldots$. This result demonstrates that the Fourier series for x can be found quite simply owing to the linearity of the equation for x.

Because it is difficult to digest this complicated formula for x, various graphical methods have been developed. One useful method is the *power spectrum* method. In this, the numbers $|x_n|^2$ are plotted against the frequencies ω_n. For example, if $f(t)$ is the Dirac delta function, then $c_n = 1$ for all n, and

$$x_n = \frac{1}{v^2 - \omega_n^2}.$$

The power spectrum shows what modes are major contributors to the signal $x(t)$; in this case they are the ones where $\omega_n^2 \approx v^2$.

We have performed a Fourier transform in finding the sequence $\{x_n\}$. This sequence is called the *generalized Fourier transform* of x and it is defined by the formulas

$$x_n = \lim_{T \to \infty} \frac{1}{T} \int_0^T x(t) \exp(-i\omega_n t) \, dt.$$

This whole business is now highly developed and automated. When a signal is read electronically, an on-line Fourier transform can be performed and the power spectrum can be plotted immediately, or it can be digitized and stored.

A.2.4 Laplace transforms: the long form

Laplace's transform provides a useful method for solving differential equations because it transforms differential equations into algebraic equations. The price for this convenience is paid when the transform variables are converted back to the original ones.

If $f(t)$ is a smooth function that does not grow faster than an exponential as $t \to \infty$, then we can define the Laplace transform of f to be

$$\tilde{f}(p) = \int_0^\infty \exp(-pt) f(t) \, dt,$$

where p is called the transform variable. It follows that the transform of the derivative of f, $\dot{f}$, is

$$\tilde{\dot{f}}(p) = p \tilde{f}(p) - f(0).$$

Laplace's transform of the differential equation

$$\dot{f} + f = 0$$

is

$$(p+1)\tilde{f}(p) = f(0).$$

Therefore, $\tilde{f}(p) = f(0)/(p+1)$.

In order to find $f(t)$ from this formula, we must use the inversion formula. This involves integration in the complex plane, and so using this method usually requires some more sophisticated techniques. However, the inversion formula is

$$f(t) = \frac{1}{2\pi i} \oint_C \exp(pt) \tilde{f}(p) \, dp,$$

where the contour C is chosen in the complex plane to enclose the singularities of $\tilde{f}$.

For the differential equation above, we have

$$f(t) = \frac{1}{2\pi i} \oint_C \exp(pt) \frac{f(0)}{p+1} \, dp,$$

where C is the straight line from $2 - i\infty$ to $2 + i\infty$. The only singularity of $\tilde{f}$ is at $p = -1$, where it has a simple pole. The method of residues then shows that this integral is $\exp(-t) f(0)$. Therefore,

$$f(t) = e^{-t} f(0).$$

A.3 Exercises

1. *Complex exponentials.* Show that there are two linearly independent solutions $x_1(t) = \exp(i\omega t)$ and $x_2(t) = \exp(-i\omega t)$ of the harmonic oscillator equation.

2. *Initial value problem.* Find what a_1 and a_2 must be so that $x(t) = a_1 x_1(t) + a_2 x_2(t)$ from Exercise 1 satisfies the initial conditions $x(0) = 1, \dot{x}(0) = 0$.

3. *Euler's formula.* Use Euler's formula [$\exp(iz) = \cos z + i \sin z$] to show that the solution found in the preceding exercises is the same as the one given in terms of $\cos \omega t$ and $\sin \omega t$.

4. *Laplace transform.* Use the Laplace transform and its inverse to solve the *RLC* circuit equation in Chapter 1.

5. *Generalized Fourier transform.* Use a generalized Fourier series to solve the equation $\ddot{x} + x = \sin 2t + \sin t\sqrt{2}$.

Appendix B

Bifurcation analysis

Consider a general network of elements that is described by a set of differential equations

$$\dot{x} = F(t, x, \Lambda)$$

where we suppose:

- The state variables $x \in E^N$ describe each element in the network. For example, in a large network of neurons, each neuron might be described by a collection of variables (membrane potentials, chemical concentrations, etc.), and x includes all of these.

- $\Lambda \in E^M$ is a set of parameters describing relevant processes in the network; for example, components of Λ might be strengths of connection from one element to another, biochemical rate constants, resistivities, capacitances, etc.

- Technical mathematical assumptions: We suppose that F is a twice continuously differentiable function taking values in E^N, specifically $F \in C^2([0, \infty) \times E^N \times E^M \to E^N)$.

    F and its first and second derivatives are uniformly bounded. The limit as $t \to \infty$, $F(\infty, x, \Lambda)$, exists as a smooth function.

Suppose this system is operating in a state x^* when $\Lambda = \Lambda^*$, so

$$\dot{x}^* = F(t, x^*, \Lambda^*).$$

An important question is "When does the stability of this state change?" It is in such cases that the system can experience a phase change, and such events are often observable in experiments. This usually occurs when the state x^* disappears and is replaced by another state, and it is detectable mathematically

191

when an eigenvalue of an associated linear system passes through the imaginary
axis from left to right.

B.1 The linear problem

Linearizing the system about x^* is accomplished by setting $x = x^* + \delta x$ and
ignoring terms involving higher powers of δx. The result is

$$\dot{\delta x} = A^*(t)\delta x,$$

where

$$A^*(t) = \frac{\partial F}{\partial x}\left(t, x^*(t), \Lambda^*\right).$$

We restrict our attention to the case where A^* is a diagonal matrix. For other
cases and additional literature, see [58]. We suppose here that

$$A^*(t) = \text{diag}\left(0, \ldots, 0, \mu_{k+1}(t), \ldots, \mu_N(t)\right)$$

and $\text{Re}\,(\mu_j(t)) \le -\delta < 0$ for $0 \le t < \infty$. This condition can be relaxed somewhat to simply requiring that as t changes the real parts of these eigenvalues
remain bounded away from the imaginary axis in the complex plane, but we do
not pursue this here.

B.2 Bifurcations (phase changes) in general networks

Indications of an imminent bifurcation are related to eigenvalues of a linear
problem approaching zero. Therefore, it is not surprising that we begin discussion of bifurcations with a discussion of linear systems. We do this in the
context of the Liapunov–Schmidt method, which we use to obtain canonical
problems.

B.2.1 Fredholm's alternative

Consider a linear problem

$$Ax = y$$

that we wish to solve for the vector x in terms of the matrix A and the forcing
(or input) vector y. If the matrix A has no zero eigenvalues, then there is a
unique solution to this problem, and it is given by the formula

$$x = A^{-1}y,$$

where the inverse of A, denoted here by A^{-1}, can be calculated in various ways (see [125]).

If A has a zero eigenvalue, as we assume above, then this method does not work, and we must try something else. For example, the problem may have no solution (e.g., $0 \cdot x = y$). However, there are conditions under which solutions exist, and there are ways to calculate them.

Suppose that A has some zero eigenvalues, as above, and let $v(A)$ denote the set of all vectors, say v, for which $Av = 0$. We write this concisely as

$$A : v(A) \to 0.$$

Let the set of all vectors that are orthogonal to all vectors in $v(A)$ be denoted by $S = v(A)^{\perp}$. The set $v(A)$ is also called the kernel of A; sometimes it is denoted by $v(A) = \ker(A)$.

Note that if there is a solution to $Ax = y$, then for any vector, say ψ, that is in $v(A^*)$ (where $*$ denotes the conjugate transpose of A) we have that the dot product

$$\psi \cdot y = \psi \cdot (Ax) = (A^* \psi) \cdot x = 0.$$

Therefore, if there is a solution, we must have $y \in v(A^*)^{\perp}$, that is, y must be orthogonal to all of the vectors in $v(A^*)$. It is known that A is a one-to-one and invertible mapping of $v(A)^{\perp}$ onto $v(A^*)^{\perp}$.

We write this observations more concisely as

$$A : v(A)^{\perp} \leftrightarrow v(A^*)^{\perp}.$$

This collection of observations is referred to as Fredholm's Alternative. If $y \in v(A^*)^{\perp}$, then there is a solution, and it has the form

$$x = c_1 \phi_1 + \cdots + c_k \phi_k + w,$$

where the vectors $\{\phi_1, \ldots, \phi_k\}$ are assumed to span the set $v(A)$, the coefficients $\{c_1, \ldots, c_k\}$ are arbitrary constants, and $w \in v(A)^{\perp}$ is uniquely determined.

B.2.2 Liapunov and Schmidt's method

Suppose that we wish to solve a nonlinear problem

$$Ax = f(x, \lambda),$$

where λ is a collection of parameters. Suppose, like we do in the implicit function theorem, that

1. there is a solution for some value of λ, say

$$Ax_0 = f(x_0, \lambda_0),$$

and

2. the function f is differentiable at and near (x_0, λ_0).

However, we do not suppose that the Jacobian matrix for this equation is invertible. Still, we would like to solve the problem for solutions x that are near x_0 when λ is near λ_0.

Using our knowledge of Fredholm's Alternative, we write a possible solution in the form

$$x = \sum_{j=1}^{k} c_j \phi_j + w,$$

where we hope to find coefficients c_j and a particular solution $w \in v(A)^\perp$ such that this is actually a solution. (Here, as before, $\{\phi_j\}$ span $v(A)$.)

For any choice of $c_1, \ldots, c_k$ and $\lambda \approx \lambda_0$, there is a unique solution for $w \in v(A)^\perp$, say $w^*(c_1, \ldots, c_k, \lambda)$. This follows from the Implicit Function Theorem. In order that the full problem be solvable, we must have that each vector $\psi_j \in v(A^*)^\perp$ satisfies the solvability condition:

$$0 = \psi_j \cdot f\left(\sum_{j=1}^{k} c_j \phi_j + w^*(c_1, \ldots, c_k, \lambda), \lambda\right) \qquad \text{(B.1)}$$

for $j = 1, \ldots, k$. If these k equations can be solved for the k unknowns $c_1, \ldots, c_k$ as functions of the parameters λ, then a solution of the original problem has been found.

Equation B.1 is the *bifurcation equation* for this problem, and the method used here to derive it is referred to as being the Liapunov–Schmidt method.

This method handles the case where we wish to find static states of the system. It is possible to extend it directly to cases where the solutions change with time: Any $v(t) \in E^N$ can be written uniquely as

$$v(t) = c_1(t)\mathbf{e}_1 + \cdots + c_k(t)\mathbf{e}_k + w_{k+1}(t)\mathbf{e}_{k+1} + \cdots + w_N(t)\mathbf{e}_N,$$

where $\mathbf{e}_j$ is the jth standard basis element:

$$\mathbf{e}_j = (\delta_{1,j}, \ldots, \delta_{i,j}, \ldots, \delta_{N,j}).$$

We can find dynamical equations for the coefficients in this decomposition following the same method. We do this next for a special case where $k = 1$.

B.3 Newton's method

We consider the case where $k = 1$ and write $c_1 = c$. This restriction is of interest here since in our networks most multiple bifurcations occur in a nearly identical way due to high redundancy in the system. Therefore, the (almost) simultaneous bifurcations occur in such a way that each can be isolated in the network. Studying one example indicates the general situation.

Setting

$$x(t) = c(t)\mathbf{e}_1 + w(t) + x^*(t),$$

where $w \cdot \mathbf{e}_1 \equiv 0$, and setting $\lambda = \Lambda - \Lambda^*$ gives

$$\dot{c} = f(t, c, w, \lambda)$$
$$= a_2(t)c^2 + a_3(t)c^3 + b_2(t)(w, w) + (m_{1,0}(t), \lambda) + (m_{1,1}, \lambda)c + h.o.t.$$

and

$$\dot{w} = B(t)w + g(t, c, w, \lambda),$$

where

$g = O(|c|^2, |c||w|, |w|^2, |\lambda|)$;
$b_2(t)(\cdot, \cdot) : E^{N-1} \times E^{N-1} \to E^1$ is a bilinear form that is continuous in t;
$m_{1,0}, m_{1,1} \in E^M$; and
$B(t) = \text{diag}(\mu_2, \dots, \mu_N), \mu_j(t) \leq \delta < 0$.

(As in the text, the notation $h.o.t.$ denotes higher order terms.)

The Weierstrass Preparation Theorem implies that smooth small solutions can be found in fractional power series of ε. This suggests rescaling the variables to pick out the relative sizes of c, λ, etc. that are the basis for constructing various solutions.

We rescale the variables c, w, and λ in powers of a parameter ε, say

$$c = \varepsilon^\alpha c_\alpha + h.o.t.,$$

$$\lambda = \varepsilon^\beta \lambda_\beta + h.o.t.,$$

$$w = \varepsilon^\gamma w_\gamma + h.o.t.,$$

and determine for which values of α, β, and γ terms in the equation for c balance.

This gives

$$\varepsilon^\alpha \dot{c}_\alpha = \varepsilon^\beta (m_{1,0}\lambda_\beta) + \varepsilon^{2\alpha} a_2(t)c_\alpha^2 + \varepsilon^{3\alpha} a_3(t)c_\alpha^3$$
$$+ (m_{1,1}, \lambda_\beta)c_\alpha \varepsilon^{\beta+\alpha} + b_2(t)(w_\gamma, w_\gamma)\varepsilon^{2\gamma} + h.o.t.$$

and

$$\varepsilon^{\gamma}\dot{w}_{\gamma} = \varepsilon^{\gamma} B(t)w_{\gamma} + \varepsilon^{2\alpha} A_2(t)c_{\alpha}^2 + \varepsilon^{\beta} M_{\beta}(t)\lambda_{\beta} + h.o.t.$$

The idea is to pick powers α and β that lead to balancing terms in the equation for c. The scaling for w follows automatically from the Implicit Function Theorem since the matrix B is invertible. This can be done using Newton's polygons. Two cases of particular interest here are:

1. For all $0 \leq t \leq \infty$, $(m_{1,0}(t), \lambda_{\beta}) \equiv 0$, $a_2(t) \equiv 0$, $a_3(t) \neq 0$ and $(m_{1,1}(t), \lambda_{\beta}) \neq 0$, in which case we have

$$\varepsilon^{\alpha}\dot{c}_{\alpha} = \varepsilon^{3\alpha} a_3(t)c_{\alpha}^3 + (m_{1,1}, \lambda_{\beta})c_{\alpha}\varepsilon^{\beta+\alpha} + h.o.t.,$$

and we take

$$2\alpha = \beta,$$

so

$$\dot{c}_{\alpha} = \varepsilon^{\beta}\left(a_3(t)c_{\alpha}^3 + (m_{1,1}, \lambda_{\beta})c_{\alpha}\right) + h.o.t.;$$

2. $(m_{1,0}(t), \lambda_{\beta}) \neq 0$, $a_2(t) \neq 0$ for all $t \geq 0$, in which case we have

$$\varepsilon^{\alpha}\dot{c}_{\alpha} = \varepsilon^{\beta}\left(m_{1,0}(t), \lambda_{\beta}\right) + \varepsilon^{2\alpha} a_2(t)c_{\alpha}^2 + h.o.t.,$$

and we take

$$2\alpha = \beta,$$

so

$$\dot{c}_{\alpha} = \varepsilon^{\alpha}\left(a_2(t)c_{\alpha}^2 + \left(m_{1,0}(t), \lambda_{\beta}\right)\right) + h.o.t.$$

(where λ_{β} is fixed here).

There are many other cases to consider; for example, these relations could change with time. However, we consider only these two cases here, and we take $\beta = 1$ and $\alpha = 1/2$.

B.4 Reduction to a canonical model

In either case, we have a singular perturbation problem that can be solved using the method of matched asymptotic expansions [39, 58]. In the second case it has the form

$$\dot{c} = \sqrt{\varepsilon} C(t, c, w, \varepsilon),$$
$$\dot{w} = B(t)w + \sqrt{\varepsilon} G(t, c, w, \varepsilon).$$

The solution has the form

$$c = c^*(\sqrt{\varepsilon} t, \sqrt{\varepsilon}) + C(t, \sqrt{\varepsilon}),$$
$$w = w^*(\sqrt{\varepsilon} t, \sqrt{\varepsilon}) + W(t, \sqrt{\varepsilon}),$$

where $|C| + |W| \leq K \exp(-\kappa t)$ on some interval depending on what c^* does.

If c^* stays bounded for all $t \geq 0$, then this result can be uniformly valid where C and W decay exponentially as $t \to \infty$.

The leading term in c^* is determined by solving the equation

$$\frac{dc^*}{ds} = \left(m_{1,0}(\infty), \lambda_\beta\right) + a_2(\infty)c^{*2},$$

where $s = \sqrt{\varepsilon} t$ describes the slow time scale.

The first case has $c^* = 0$ is a solution that is stable under persistent disturbances if $a_3(\infty) < 0$ and $(m_{1,1}(\infty), \lambda_\beta) \leq 0$, and $c^* = \pm\sqrt{(m_{1,1}, \lambda_\beta)}$, otherwise. But in the second, the behavior is more complicated [130, 111].

The first case corresponds to the occurrence of a cusp (or pitchfork) bifurcation, and it entails a corresponding exchange of stabilities. Its analysis is complete. In the second, the system encounters a fold (or saddle-node) bifurcation.

B.5 Saddle-node bifurcation

Let us examine the saddle-node case. First, we consider the canonical equation

$$\dot{c} = a + bc^2.$$

Setting

$$c = \tan\left(\frac{\theta}{2}\right)$$

we obtain

$$\dot{\theta} = \omega + \zeta \cos \theta,$$

where $\omega = a + b$ and $\zeta = a - b$. This equation's dynamics are described by the diagram

for $0 \leq \omega/\zeta < 1$ and $\omega/\zeta > 1$. Here 'o' denotes an unstable node and '•' denotes a stable node.

This model is a valid approximation to solutions of the system only for $|\theta| \ll 1$. We must know more about the global behavior of solutions since the canonical problem only indicates that in some cases the solution wanders off, possibly going to another quasi-static manifold.

Because of this, it is useful to consider a *saddle-node bifurcation on a limit cycle*, as suggested by the canonical model, namely,

$$\dot\theta = \omega + \cos\theta.$$

As ω increases through the value $\omega = 1$, the solutions pass from a saddle-node configuration to a limit cycle. The relation of this to general systems is described in the preface.

B.5.1 Weakly connected networks

Izhikevich [74] has shown that for weakly connected networks like

$$\dot x_j = F_j(x_j) + \varepsilon G_j(x)$$

near a saddle-node bifurcation, the canonical model is

$$\dot\theta_j = \omega_j + \cos\theta_j + \sum_{i\neq j} C_{i,j}\cos(\theta_j - \theta_i + \psi_{i,j}),$$

where the matrix $C_{i,j}$ describes connections strengths from site i to site j and $\psi_{i,j}$ describes the connection polarity.

B.5.2 VCON

The VCON model is

$$\dot\theta = a,$$

$$\tau\dot a + a = \omega + \cos\theta + P(\theta, \Theta).$$

Here a is the firing rate or activity of the oscillator, Θ is the phase of the input variable, θ is the oscillator's phase. The activity a describes release of neuro-transmitter, which is proportional to firing rate, and P describes the membrane potential and synaptic input.

This model is quite similar to the canonical model for a saddle-node on a limit cycle bifurcation. These models are studied in this text. (See Figure 3.14.)

B.5.3 Pendulum bifurcations

A canonical model relevant to our work is illustrated by a pendulum having applied torque and friction, which can be described by the equation

$$\ddot{\theta} + r\dot{\theta} + \sin\theta = \omega.$$

This model can have three possible solutions:

1. An isolated globally stable periodic solution (i.e., a closed curve on the cylinder θ MOD 2π), so there is a solution with $\theta(t + T) = \theta(t) + 2\pi$ for all t. This motion corresponds to the pendulum clocking around its support point.
2. A saddle point and a stable node. This corresponds to the pendulum coming to rest at a balance between the applied torque, the dissipation, and gravity.
3. A single stable periodic solution, like θ^* above, coexisting with a saddle and a stable node. There are two possible stable motions in this case: clocking of the pendulum and a balanced equilibrium.

Such bifurcations as these can be described as being saddle-node bifurcations near a limit cycle. The complication here is that both the saddle-node structure and a stable limit cycle can coexist for intervals of parameter values. The VCON model is quite similar to this. See [124, 91].

B.6 Other bifurcations

Two other prominent bifurcations have been extensively studied in the context of neurobiology. These are the pitchfork and the Andronov–Hopf bifurcations. The following two situations are observed in applications.

B.6.1 Pitchfork bifurcation

When the canonical problem has the form

$$\dot{c} = ac + bc^3$$

the problem is somewhat easier since the solutions remain near 0 when a and b are near zero. To fix ideas, suppose that $b < 0$. By rescaling the problem,

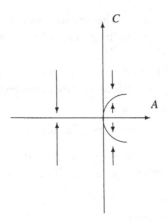

Figure B.1. A pitchfork bifurcation.

we get

$$\dot{C} = AC - C^3.$$

As A increases through 0, the solution $C = 0$ becomes unstable and two new solutions appear: $C = \pm\sqrt{A}$. These are stable. When these solutions are plotted on $C(\infty)$ versus A, the diagram resembles a pitchfork, as shown in Figure B.1.

B.6.2 Hopf bifurcations

We described the appearance of periodic solutions through bifurcations in our discussion of the FitzHugh–Nagumo model. In this, as a parameter is increased, a stable node becomes unstable, and at that point a new stable oscillation appears to grow from the node.

This phenomenon is illustrated by the following model, which we present already converted to polar coordinates:

$$\dot{r} = r(a - r^2),$$

$$\dot{\theta} = \omega.$$

As the parameter a is increased through the value $a = 0$, the node $r = 0$ changes from being stable to being unstable, and a new static state for r (but not static for the system) appears, namely $r = \sqrt{a}$. There is also an antiphase solution $r = -\sqrt{a}$. In terms of physical variables, we would have

$$x = a\cos\theta,$$

$$y = a\sin\theta.$$

Such problems can usually be converted to consideration of nonoscillatory bifurcations by converting it to phase-amplitude coordinates, as was done here using polar coordinates, and considering the single equation

$$\frac{dr}{d\theta} = \frac{r(a - r^2)}{\omega},$$

which encounters a pitchfork, or cusp, bifurcation at $a = 0$.

Hopf bifurcations can occur when a pair of eigenvalues of the linear problem pass through the imaginary axis, but not at the origin. Usually, an oscillation splits from the static state in this case.

The following example, referred to as being a $\lambda - \omega$ system, is typical of this:

$$\begin{pmatrix} \dot{x} \\ \dot{y} \end{pmatrix} = \begin{pmatrix} A - x^2 - y^2 & -\omega \\ \omega & A - x^2 - y^2 \end{pmatrix} \begin{pmatrix} x \\ y \end{pmatrix}.$$

Introducing polar coordinates

$$r^2 = x^2 + y^2,$$

$$\tan \theta = y/x,$$

we get

$$\dot{r} = r(A - r^2),$$

$$\dot{\theta} = \omega.$$

Finally, introducing θ as a new time variable, we get

$$\frac{dr}{d\theta} = \frac{r(A - r^2)}{\omega}.$$

As A increases through zero, we have a pitchfork bifurcation for the amplitude (r) of an oscillation. For $A < 0$, $r = 0$ is stable, and for $A > 0$, $r \to \pm\sqrt{A}$ if $r(0) \neq 0$.

B.7 Summary

In this appendix we have seen that:

- Near a bifurcation the general system can be reduced to a canonical model using singular perturbation methods.
- The canonical model, although useful only locally, motivates consideration of a more special model whose global dynamics are known.
- VCON arises as a natural model in the sense that it includes the canonical model of a saddle-node on a limit cycle bifurcation.

- Like the pendulum model, the VCON model can be analyzed for coexisting saddle nodes and oscillations.

- As the bifurcation point is approached from above, a SNLC oscillation will approach a fixed amplitude but decreasing frequency. A Hopf oscillation will approach a fixed frequency but with decreasing amplitude. These features are observed in experiments.

References

[1] M. Abramowitz and I. A. Stegun, *Tables of Mathematical Functions*. New York: Dover. 1970.
[2] M. A. Arbib, eds. *Handbook of Brain Science*. New York: Springer-Verlag. 1982.
[3] J. A. Anderson and E. Rosenfeld (1988), *Neurocomputing: Foundations of Research*. Cambridge, MA: MIT Press.
[4] V. I. Arnold (1984), *Mathematical Methods of Classical Mechanics*. New York: Springer-Verlag.
[5] C. Ascoli, M. Barbi, S. Schillemi, and D. Petracchi (1977), Phase-locked responses in the Limulus lateral eye, Theoretical and experimental investigation, *Biophysical J.* 19:219–40.
[6] F. Berthommier, J. Demongeot, and J. L. Schwartz (1989), A neural net for processing of stationary signals in the auditory system, *IEE Proc. (London)*.
[7] F. Berthommier (1989), Un modele de la relation entre tonotopie et synchronisation dans le systeme auditif. *C. R. Acad. Sci. Paris* 309:695–701.
[8] N. N. Bogoliuboff (1964), On quasiperiodic solutions in nonlinear problems in mechanics, *First Math Summer School, Kanaev Akad. Nauk Ukr. SSR* 79–80.
[9] R. Borisyuk, G. Borisyuk, et al. Personal Communication.
[10] G. N. Borisyuk, R. M. Borisyuk, A. B. Kirillov, V. I. Kryukov, and W. Singer, Modeling of oscillatory activity of neuron assemblies of the visual cortex. Preprint.
[11] *The Brain: A Users Manual*, Berkeley, New York, 1980.
[12] D. Bramble and D. R. Carrier (1983), Running and breathing in mammals, *Science* 21:251–56.
[13] H. Carrillo (1983), *The method of averaging and stability under persistent disturbances with applications to phase locking*, Dissertation, U. Nat. Aut. Mex.
[14] G. Carpenter and S. Grossberg (1983), A neural theory of circadian rhythms: the gated pacemaker, *Biol. Cybernetics* 4:35–39.
[15] A. N. Chetaev (1985), *Neural Nets and Markov's chain*. Mockba: NAUKA.
[16] M. A. Cohen and S. Grossberg (1983), Absolute stability of global pattern formation and parallel memory storage by competitive neural networks, *IEEE Trans.* SMC 13:815–26.
[17] D. S. Cohen, F. C. Hoppensteadt, and R. M. Miura (1977), Slowly modulated oscillations in nonlinear diffusion processes, *SIAM J. Appl. Math.* 33:217–29.
[18] B. W. Connors and M. J. Gutnick (1990), Intrinsic firing patterns of diverse neocortical neurons, *Trends in Neural Sciences* 13:99–100.

203

[19] J. Conway (1970), in *The Game of Life, in Scientific American*, M. Gardner, ed.

[20] H. S. M. Coxeter (1961), *Introduction to Geometry*, New York: Wiley.

[21] F. Crick (1984), Function of the thalamic reticular complex: the searchlight hypothesis, *Proc. Natl. Acad. Sci. (USA)* 81:4586–90.

[22] A. Denjoy (1932), Sur les courbes definies par les equations differentielles a la surface du tor, *J. Math. Pures Appl.* 9:333–75.

[23] G. B. Ermentrout and N. Kopell (1986), Parabolic bursting in an excitable system of coupled oscillators with a slow oscillation, *SIAM J. Appl. Math.* 46:233–53.

[24] C. von Euler (1980), Central pattern generation during breathing, *Trends in Neural Sciences*, Nov.: 275–77.

[25] W. Feller (1978), *Introduction to Probability Theory and its Applications*. New York: Wiley.

[26] R. Feynmann, R. Leighton, and M. Sands (1977), *The Feynman Lectures on Physics*. Menlo Park, CA: Addison-Wesley.

[27] R. FitzHugh, Mathematical models of excitation and propagation in nerve, in *Biological Engineering*, H. P. Schwan, ed., pp. 1–85. New York: McGraw-Hill, 1957.

[28] J. E. Flaherty and F. C. Hoppensteadt (1978), Frequency entrainment of a forced van der Pol oscillator, *Stud. Appl. Math.* 58:5–15.

[29] W. J. Freeman (1961), Nervous control of shivering: Role of the fields of forel in shivering, Arctic Aeromedical Lab., Fort Wainwright, Alaska, Tech. Rep. 60-27.

[30] D. G. Stuart, W. J. Freeman, and A. Hemingway (1962), Effects of decerebration and decortication on shivering in the cat, *Neurology* 12:99–107.

[31] W. J. Freeman (1960), Nervous control of shivering: Further observations on brainstem unit potentials during shivering, Alaskan Air Command, Arctic Aeormedical Lab., Ladd AFB, Tech. Rep. 58-4.

[32] W. J. Freeman and D. D. Davis (1959), Effects on cats of conductive hypothalamic cooling, *Am. J. Physiol.* 197:145–48.

[33] S. Freud (1965), *The Interpretation of Dreams*. New York: Avon Books.

[34] W. O. Friesen and R. J. Wyman (1980), Analysis of drosophila motor neuron activity patterns with neural analogs, *Biol. Cybernetics* 38:41–50.

[35] K. Garcia-Reimbert (1984), *Stable synchronization waves in neural networks and traveling waves in glassy polymers*, Dissertation, Univ. Utah.

[36] Peter Gay (1988), *Freud: A Life for Our Time*. New York: Norton.

[37] J. O. Hirschfeleder and J. C. Giddings, Flame Properties and the Kinetics of Chain branching Reactions, 199–212, 6th Combustion Institute, *Am. Chem. Soc.*, 1956.

[38] L. Glass and M. Mackey (1979), A simple model for phase locking of biological oscillators, *J. Math. Biol.* 7:339–52.

[39] N. Gordon and F. C. Hoppensteadt (1975), Nonlinear stability analysis of static states which arise through bifurcation, *Comm. Pure Appl. Math.* 28:355–73.

[40] Gray's Anatomy (1995), 38th edition, Churchill Livingstone, Edinburgh.

[41] J. M. Greenberg and S. P. Hastings (1978), Spatial patterns for discrete models of diffusion in excitable media, *SIAM J. Appl. Math.* 34:515–23.

[42] S. Grossberg (1988), Nonlinear neural networks: Principles, mechanisms and architectures, *Neural Networks* 1:17–61.

[43] R. Guttman, L. Feldman, and E. Jacobsson (1980), Frequency entrainment of squid axon membrane, *J. Membrane Biol.* 56:9–18.

[44] P. J. Hagan, Frequency locking in Josephson point contacts, in *Coupled Nonlinear Oscillators*, J. Chandra and A. C. Scott, eds., pp. 31–41. New York: North-Holland.

[45] W. Hahn (1967), *Stability of Motion*. New York: Springer-Verlag.

[46] F. Ratliff, H. K. Hartline, W. H. Miller (1963), Spatial and temporal aspects of retinal inhibitory interactions, *J. Optical Soc. Am.* 53:110–20.

[47] D. Hebb (1949), *The Organization of Behavior.* New York: Wiley.

[48] H. L. F. von Helmholtz, see [36].

[49] A. V. Hill, B. Katz, and D. Y. Solandt (1936), *Proc. R. Soc. London* B121:74.

[50] A. L. Hodgkin (1964), *The Conduction of the Nervous Impulse.* Liverpool: Liverpool Univ. Press.

[51] A. L. Hodgkin and A. F. Huxley (1952), Currents carried by sodium and potassium ions through the membrane of the giant axon of Loligo, *J. Physiol.* 117:500–44.

[52] A. V. Holden (1976), The response of excitable membrane models to cyclic input, *Biol. Cybernetics* 21:1–8.

[53] J. J. Hopfield (1982), Neuronal networks and physical systems with emergent collective computational abilities, *Proc. Natl. Acad. Sci. (USA)* 79:2554–58.

[54] F. C. Hoppensteadt and J. P. Keener (1982), Phase locking in biological clocks, *J. Math. Biol.* 15:339–49.

[55] F. C. Hoppensteadt (1994), Simulation of tonotopic neural circuit, in *Mathematics Applied to Biology and Medicine*, J. Demongeot and V. Capasso, eds., pp. 125–29. Winnipeg: Wuerz.

[56] F. C. Hoppensteadt and C. S. Peskin (1992), *Mathematics of Medicine and the Life Sciences.* New York: Springer-Verlag.

[57] F. C. Hoppensteadt (1986), *Introduction to the Mathematics of Neurons.* Cambridge: Cambridge Univ. Press.

[58] F. C. Hoppensteadt (1993), *Analysis and Simulation of Chaotic Systems.* New York: Springer-Verlag.

[59] F. C. Hoppensteadt (1992), Signal processing by model neural networks, *SIAM Review* 34:426–44.

[60] F. C. Hoppensteadt (1989), Intermittent chaos, self organization and learning from synchronous synaptic activity in model neuron networks, *Proc. Natl. Acad. Sci. (USA)* 86:2991–95.

[61] F. C. Hoppensteadt (1986), Analysis of a VCON neuromime, in *Lect. Notes in Biomathematics*, vol. 66, pp. 150–59. New York: Springer-Verlag.

[62] F. C. Hoppensteadt, H. S. Salehi, and A. V. Skorokhod (1995), Randomly perturbed Volterra integral equations and some applications, *Stochastics Stochastic Rep.* 54:89–125.

[63] F. C. Hoppensteadt (1989), Neural prisms, *Proc. 32nd Circuits Conference*, IEEE.

[64] F. C. Hoppensteadt (1991), The searchlight hypothesis, *J. Math. Biol.* 29:689–91.

[65] F. C. Hoppensteadt (1979), Lectures on Biological Rhythms, University of Utah Lecture Notes.

[66] F. C. Hoppensteadt (1982), *Mathematical Methods of Population Biology*, Cambridge: Cambridge Univ. Press.

[67] F. C. Hoppensteadt and E. Izhikevich (1996), *Biological Cybernetics.* 75:117–135.

[68] P. Horowitz and W. Hill (1989), *The Art of Electronics*, 2nd ed. Cambridge: Cambridge Univ. Press.

[69] D. H. Hubel and T. N. Wiesel (1972), Laminar and columnar distribution of geniculocortical fibers in the macaque monkey, *J. Comp. Neurol.* 146:421–50.

[70] D. H. Hubel and T. N. Wiesel (1979), Brain mechanisms of vision, in *The Brain, A Scientific American Book.*

[71] J. Huygens (1665), *Oeuvres Completes*, La Societe Hollandaise des Sciences, M. Nijhoff, La Haye 5 (1893) 243–56.

[72] *The I Ching* or *The Book of Changes* (1950), interpreted and translated by R. Wilhelm and C. F. Baynes, Bollingen Series. Princeton, NJ: Princeton Univ. Press.

[73] Nonlinear phenomena in power systems (1995), special issue *Proc. IEEE* 83:1439–594.

[74] E. Izhikevich (1996), *Bifurcations in brain dynamics*, Ph.D. Dissertation, Dept. Math., Michigan State Univ.

[75] E. Izhikevich and F. Hoppensteadt (in press), *Weakly Connected Oscillatory Neural Networks*. New York: Springer-Verlag.

[76] E. R. John (1967), *Mechanisms of Memory*. New York: Academic Press.

[77] C. G. Jung (1928), On psychic energy, (in *Collected Works*, vol. 8, 1969); Synchronicity, Bollingen Series, Princeton, NJ: Princeton Univ. Press, 1974; W. Pauli (1955), *The Interpretation and Nature of the Psyche*, Bollingen Series. New York: Pantheon.

[78] A. E. Kammer (1968), Motor patterns during flight and warm-up in lepidoptera, *J. Exp. Biol.* 48:89–109.

[79] A. E. Kammer (1971), The motor output during turning flight in a hawkmoth, Manduca sexta, *J. Insect. Physiol.* 17:1073–86.

[80] A. E. Kammer and S. C. Kinnamon (1979), Maturation of the flight motor pattern without movement in Manduca sexta, *J. Comp. Physiol.* 130:29–37.

[81] E. R. Kandel, J. H. Schwartz, and T. M. Jessell (1991), *Principles of Neural Science*, 3rd ed. New York: Elsevier.

[82] K. M. Harris, S. B. Kater, Dendritic spines, *Annu. Rev. Neurosci.* 17 (1994) 341–71.

[83] B. Katz (1966), *Nerve, Muscle and Synapse*. New York: McGraw-Hill.

[84] J. P. Keener (1986), Spiral waves in excitable media, in *Lect. Notes in Biomathematics*, vol. 66, pp. 115–27. New York: Springer-Verlag.

[85] J. P. Keener (1982), Operational amplifier rendition of the FitzHugh-Nagumo circuit. Preprint.

[86] W. O. Kermack and A. G. McKendrick (1927), A contribution to the theory of epidemics, *Proc. R. Soc. London* A 115:700–21.

[87] B. W. Knight (1972), Dynamics of encoding in a population of neurons, *J. Gen. Physiol.* 59:734–66.

[88] S. W. Kuffler and J. G. Nicholls (1984), *From Neuron to Brain*, 2nd ed. Sunderland, MA: Sinauer.

[89] Y. Kuramoto (1975), Self-entrainment of a population of coupled nonlinear oscillators, in *Int. Symp. Math. Problems in Theoretical Physics*, H. Araki, ed., Lect. Theor. Physics, vol. 39. New York: Springer-Verlag.

[90] L. Lapicque (1907), Recherches quantitatives sur l'excitation electriques des nerfs traitee comme une polarization, *J. Physiol. Pathol. Gen.* 9:620–35.

[91] M. Levi, F. Hoppensteadt, and W. L. Miranker (1978), Dynamics of the Josepshon junction, *Quart. J. Appl. Math.*, July: 167–90.

[92] R. J. MacGregor and E. Lewis (1977), *Neural Modelling*. New York: Plenum Press.

[93] H. Antosiewicz (1958), A survey of Liapunov's second method, in *Contributions to the Theory of Nonlinear Oscillations*, vol. IV, S. Lefschetz, ed. Princeton, NJ: Princeton Univ. Press.

[94] W. C. Lindsey (1972), *Synchronous Systems in Communications and Control*. Englewood Cliffs, NJ: Prentice-Hall.

[95] M. Steriade, E. G. Jones, and R. R. Llinas (1990), *Thalamic Oscillations and Signalling*. New York: Wiley.

[96] W. S. McCulloch and W. S. Pitts (1943), A logical calculus of the ideas immanent in nervous activity, *Bull. Math. Biophysics*, 5:115–33.

[97] I. G. Malkin (1959), *Theorie der Stabilitaet einer Bewegung*. Munich: Oldenbourg.

[98] K. D. Miller, J. B. Keller, and M. P. Stryker (1989), Ocular dominance column development: analysis and simulation, *Science* 245:605–15.

[99] H. C. Tuckwell and R. M. Miura (1978), A mathematical model for spreading cortical depression, *Biophys. J.* 23:257–76.

[100] G. K. Moe, W. C. Rheinboldt, and J. A. Abildskov (1964), A computer model of atrial fibrillation, *Am. Heart J.* 67:200–20.

[101] G. F. Newell and E. Montroll (1953), On the theory of the Ising model of ferromagnetism, *Rev. Modern Physics* 25:353–89.

[102] J. Moser (1966), On the theory of quasiperiodic motions, *SIAM Rev.* 8:145–71.

[103] A. V. Oppenheim and R. W. Shaeffer (1975), *Digital Signal Processing.* Englewood Cliffs, NJ: Prentice-Hall.

[104] S. L. Palay and V. C. Palay (1974), *Cerebral Cortex*, New York: Springer-Verlag.

[105] T. N. Parks and E. W. Rubel (1975), Organization and development of brainstem auditory nuclei of the chicken, *J. Comp. Neurology* 164:435–48.

[106] H. D. Patton et al., (1989), *Textbook of Physiology.* Philadelphia: Saunders.

[107] R. Penrose (1994), *Shadows of the Mind.* Oxford: Oxford Univ. Press.

[108] D. H. Perkel, J. Schulman, T. Bullock, G. Moore, and J. Segundo (1964), Pacemaker neurons: Effects of regularly spaced synaptic input, *Science* 145: 61–63.

[109] D. Pettigrew (1994), *Simulation of ocular dominance in the primary visual cortex using Voltage Controlled Oscillator Neurons*, MS Thesis, Mich. State Univ.

[110] B. Platt (1983), Periodic patterns in cellular automata. MS Thesis, Univ. Utah.

[111] L. S. Pontryagin (1961), Asymptotic behavior of the solutions of systems of differential equations with a small parameter in the higher derivatives, *AMS Transl. Ser.* 2, 18:295–320.
   E. F. Mishchenko (1961), Asymptotic calculation of periodic solutions of systems of differential equations containing small parameters in the derivatives, *AMS Transl. Ser.* 2, 18:199–230.

[112] W. Rall (1962), Theory of physiological properties of dendrites, *Ann. NY Acad. Sci.* 96:1071–92.

[113] W. Rall (1977), Core conductor theory and cable properties of neurons, in *Handbook of Physiology – The Nervous System I.* Bethesda: Am. Physiol. Soc.

[114] A. Resigno, R. B. Stein, R. I. Purple, and R. E. Poppele (1972), A neuronal model for the discharge patterns produced by cyclic inputs, *Bull. Math. Biophysics* 32:337–53.

[115] F. M. A. Salam, J. E. Marsden, and P. P. Varaiya (1984), Arnold diffusion in the swing equations of a power system, *IEEE Trans. Circuits and Systems*, CAS-31, No. 8, Aug.: 673–88.

[116] A. Scott (1970), *Active and Nonlinear Wave Propagation in Electronics*, New York: Wiley-Interscience.

[117] A. Selversten and P. Mazzoni (1989), Flexibility of computational units in invertebrate cpg's, in *The Computing Neuron*, R. Durbin, ed. Wolkingham, England: Addison-Wesley.

[118] N. N. Semenov (1935), *Chemical Kinetics and Chain Reactions.* Oxford: Clarendon.

[119] G. M. Shepherd (1983), *Neurobiology.* Oxford: Oxford U. Press.

[120] J. Si (1995), Analysis and synthesis of a class of discrete-time neural networks with multi-level threshold neurons, *IEEE Trans. on Neural Networks* 6:1–12.

[121] W. E. Skaggs and B. McNaughton (1996), Replay of neuronal firing sequences in rat hippocampus during sleep following spatial experiences, *Science* 271:1870–73.

[122] A. Skorokhod, H. Salehi, and F. C. Hoppensteadt. Asymptotic and Ergodic Analysis of Markov Chains in Random Environments. In preparation.

[123] J. Smoller (1983), *Shock Waves and Reaction-Diffusion Equations*. New York: Springer-Verlag.

[124] J. J. Stokers (1951), *Nonlinear Vibrations*. New York: Interscience.

[125] G. Strang (1980), *Linear Algebra and its Applications*, New York: Academic Press.

[126] B. L. Strehler and R. Lestienne (1986), Evidence of precise time-coded symbols and memory of patterns in monkey cortical neuronal spike trains, *Proc. Natl. Acad. Sci. (USA)* 83:9812–16.

[127] S. H. Strogatz (1993), *Norbert Wiener's Brain Waves*, Biomathematics, vol. 100. New York: Springer-Verlag.

[128] D. G. Stuart, R. George, W. J. Freeman, A. Hemingway, and W. M. Price (1961), Effects of anti- and pseudo-Parkinson drugs on shivering, *Exp. Neurology* 4:106–14.

[129] R. Thom (1972), Stabilite Structurelle et Morphogenese, New York: Benjamin.

[130] R. Thom (1973), Topological models in biology, in *Towards a Theoretical Biology 3*, C. Waddington, ed. Edinburgh: Univ. Edinburgh Press.

[131] B. J. Travis (1988), A layered network of sensory cortex, in *Computer Simulation in Brain Science*, R.M.J. Cotterill, ed. Cambridge: Cambridge Univ. Press.

[132] H. C. Tuckwell (1988), *Introduction to Theoretical Neurobiology: Volume I Linear Cable Theory and Dendritic Structure*. Cambridge: Cambridge Univ. Press.

[133] A. Turing (1936), *Proc. R. Soc.* 5:230–65.

[134] B. van der Pol (1926), On relaxation oscillators, *Philos. Magazine* 2:978.

[135] Th. von Karman (1940). The engineer grapples with nonlinear problems, *Bull. Am. Math. Soc.* 615–83.

[136] A. D. Wentzel and M. I. Friedlin (1970), *Asymptotic Methods for Stochastic Differential Equations*. Moscow: MIR.

[137] N. Wiener and A. Rosenblueth (1946), The mathematical formulation of the problem of conduction of impulses in a network of connected excitable elements, specifically in cardiac muscle, *Arch. Inst. Cardiologia Mexico*, 16:205–65.

[138] N. Wiener (1961), *Cybernetics*, 2nd ed. Cambridge, MA: MIT Press.

[139] N. Wiener (1954), *The Fourier Integral and Certain of Its Applications*. New York: Dover.

[140] H. R. Wilson and J. D. Cowan (1973), A mathematical theory of the functional dynamics of cortical and thalamic nervous tissue, *Kybernetik* 13:55–80.

[141] H. R. Wilson and J. D. Cowan (1972), Excitatory and inhibitory interactions in localized populations of model neurons, *Biophysical J.* 12:1–24.

[142] A. Winfree (1980), *The Geometry of Biological Time*. New York: Springer-Verlag.

[143] D. Young (1989), *Nerve Cells and Animal Behavior*, Ch. 5. Cambridge: Cambridge Univ. Press.

Index

Printed in the United States
By Bookmasters